Foreword

Remember the initial excitement of experimenting with Lego robotics? It felt like the entire universe was open to you! With so many robots to build to tackle interesting challenges, perform experiments, and make interesting algorithms, the possibilities were endless. From modifying some bricks with LEDs, to gluing assemblies together to make your robot better, robot building with Legos is an exciting experience.

One of my personal favorite experiments with Lego robotics was conducting an experiment to determine if using a light sensor or touch sensor (which was calibrated with springs using Hooke's law) would be better for counting pills. Between the real world applications, as well as the scientific method applied behind it, it ended up winning two gold medals at my high school and regional science fairs.

In order to accelerate towards more advanced experiments in robotics, using the RCX 2.0 was not a viable solution. The lack of ability to interface to various sensors and actuators was a major roadblock, so I adventured into the land of Arduino! Since then, my mind has been opened up to even more interesting areas of robotics.

The journey of learning about robotics is interesting and never-ending. This book will be a huge help to springboard you into more in-depth robotics. Not only will you be able to make more sophisticated robots that tackle more complex challenges or perform more elaborate experiments, but you will have a better opportunity to discover which area of robotics you enjoy the most, whether it's programming, electrical, mechanical, or design and implementation.

As you embark on your new Lego robots that utilize Arduino, it will feel like the entire universe is open to you again, with even more galaxies to explore! Embrace this, as you never know where your next robotic creation may take you. Maybe it will be like RoboBrrd, from a prototype to a kit, inspiring even more robot builders. Be sure to share your excitement for robots with your friends or an online community. Remember, robots teach us about ourselves, how we think, behave, and act.

Welcome to the awesome world of robotics! Now get on with the learning and advance your robots to the next step!

— Erin "RobotGrrl" Kennedy

Make: Lego and Arduino Projects

John Baichtal, Matthew Beckler & Adam Wolf

MAKER MEDIA

Sebastopol

Make: Lego and Arduino Projects

by John Baichtal, Matthew Beckler & Adam Wolf

Published by Maker Media, Inc., 1005 Gravenstein Highway North, Sebastopol, CA 95472.

Maker Media books may be purchased for educational, business, or sales promotional use. Online editions are also available for most titles (*my.safaribooksonline.com*). For more information, contact our corporate/institutional sales department: 800-998-9938 or *corporate@oreilly.com*.

Editors: Dale Dougherty and Brian Jepson	**Cover Designers:** Mark Paglietti and Randy Comer
Development Editor: Brian Jepson	**Interior Designer:** Ron Bilodeau
Production Editor: Kristen Borg	**Production Services:** Peter Amirault
Proofreader: nSight, Inc.	**Illustrators:** Matthew Beckler, Rebecca Demarest
Indexer: WordCo Indexing Services	**Cover Photographer:** Adam Wolf

December 2012:   First Edition.

Revision History for the 1st Edition:

2012-11-14   First release
2013-02-22   Second release

See *http://oreilly.com/catalog/errata.csp?isbn=0636920024316* for release details.

ISBN: 978-1-449-32106-2
[TI]

Contents

Preface

In the classic swords & sorcery flick Conan the Barbarian, the title character is asked "What is best in life?" To which the mighty-thewed warrior responds, "To crush your enemy, see them driven before you, and hear the lamentation of their women."

Wrong, Conan. You are wrong.

First of all, crushing people and getting off on their significant others' lamentations is mega uncool. No, Conan, what is best in life is having your cake and eating it too. Why, barbarian, should you—oh, for example—be able to play with Arduinos but not be able to integrate them into your Lego Mindstorms NXT projects?

If you don't know, the Arduino phenomenon is pretty much the coolest thing to hit the amateur electronics scene. It makes programming your own microcontroller absurdly easy. You may also be unaware that the Lego Mindstorms NXT phenomenon is pretty much the coolest thing to hit the amateur robotics scene. It makes building your own robot absurdly easy.

Make: Lego and Arduino Projects is dedicated to combining the two.

Assumptions This Book Makes

This book starts easy, with the first two projects being relatively simple and accessible to all skill levels. That said, you may find certain topics in the later chapters rather challenging. Don't worry! Everything you need in order to build the various projects is clearly explained; most of the really hard stuff is purely informational.

Additionally, we made enthusiastic use of our extensive Lego collection in designing the book's robots. If you have only a modest collection—at the very least you'll want the Lego Mindstorms NXT 2.0 set—you'll definitely need to buy more parts to build the projects.

Contents of This Book

We cover a wide variety of topics in this book. Here's what you get:

Chapter One introduces the reader to the mysteries of Lego and Arduino with our Drawbot project. It's a simple project that creates a robot that draws shapes with a pen. It'll help you delve into the mysteries of controlling Mindstorms motors with an Arduino. We also introduce the Bricktronics Shield, our Arduino add-on that interfaces between Lego and Arduino.

In Chapter Two we brush up on our Lego skills, learning about the Mindstorms set and compatible product lines like Technic and Power Functions.

Chapter Three covers the opposite side of the equation: learning about Arduino and their add-on boards, called shields.

Our second project, a Clock, debuts in Chapter Four, and will teach us how to use Mindstorms motors to precisely control the clock's hands.

Chapter Five details one of the most complicated robots in the book, a Chocolate Milk Maker that pumps milk and chocolate syrup into a cup and mixes them up with a motorized spoon.

We learn more about the mysteries of electronics in Chapter Six, discovering how electricity works, what all those components do, and how to use the electronic hobbyist's primary diagnostic tool, a multimeter.

Chapter Seven serves up another convoluted robot, a Gripperbot: a wheeled manipulator controlled with a pair of Arduino-equipped wristbands and Wii Nunchuk controllers.

Chapter Eight shows you how to build a cool electronic instrument, a Keytar. It's a Lego guitar equipped with buttons and knobs that create a variety of electronic noises!

For the final project, Chapter Nine describes our Lamp project, which details the creation of a smartphone-controlled Lego light fixture.

The book closes with Chapter Ten, where we cover a number of advanced electronic topics like choosing motors and power supplies, and how to create breadboard equivalents to our Bricktronics Shield.

Good luck and have fun!

Conventions Used in This Book

The following typographical conventions are used in this book:

Italic
> Indicates new terms, URLs, email addresses, filenames, and file extensions.

`Constant width`
> Used for program listings, as well as within paragraphs to refer to program elements such as variable or function names, databases, data types, environment variables, statements, and keywords.

`Constant width bold`
> Shows commands or other text that should be typed literally by the user.

`Constant width italic`
> Shows text that should be replaced with user-supplied values or by values determined by context.

> *This box signifies a tip, suggestion, or general note.*

> **Warning**
>
> *This box indicates a warning or caution.*

Lego CAD Conventions

In illustrating the Lego build steps throughout this book, we'll be coloring the elements that are added in each step, to distinguish them from elements added in previous steps. Look for the bright greenish color (see Figure P-1) to see the new parts added in that step.

Figure P-1 *To see what new parts are added in each step, look for the yellowish-green elements*

Using Code Examples

This book is here to help you get your job done. In general, you may use the code in this book in your programs and documentation. You do not need to contact us for permission unless you're reproducing a significant portion of the code. For example, writing a program that uses several chunks of code from this book does not require permission. Selling or distributing a CD-ROM of examples from MAKE's books does require permission. Answering a question by citing this book and quoting example code does not require permission. Incorporating a significant amount of example code from this book into your product's documentation does require permission.

We appreciate, but do not require, attribution. An attribution usually includes the title, author, publisher, and ISBN. For example: "*Make: Lego and Arduino Projects* by John Baichtal, Matthew Beckler, and Adam Wolf (MAKE). Copyright 2013 John Baichtal, Matthew Beckler, and Adam Wolf, 978-1-4493-2106-2."

If you feel your use of code examples falls outside fair use or the permission given above, feel free to contact us at *permissions@oreilly.com*.

Safari® Books Online

Safari Books Online (*http://my.safaribooksonline.com*) is an on-demand digital library that delivers expert *content* in both book and video form from the world's leading authors in technology and business. Technology professionals, software developers, web designers, and business and creative professionals use Safari Books Online as their primary resource for research, problem solving, learning, and certification training.

Safari Books Online offers a range of *product mixes* and pricing programs for *organizations, government agencies,* and *individuals.* Subscribers have access to thousands of books, training videos, and prepublication manuscripts in one fully searchable database from publishers like O'Reilly Media, Prentice Hall Professional, Addison-Wesley Professional, Microsoft Press, Sams, Que, Peachpit Press, Focal Press, Cisco Press, John Wiley & Sons, Syngress, Morgan Kaufmann, IBM Redbooks, Packt, Adobe Press, FT Press, Apress, Manning, New Riders, McGraw-Hill, Jones & Bartlett, Course Technology, and dozens *more.* For more information about Safari Books Online, please visit us *online.*

How to Contact Us

Please address comments and questions concerning this book to the publisher:

Maker Media, Inc.
1005 Gravenstein Highway North
Sebastopol, CA 95472
800-998-9938 (in the United States or Canada)
707-829-0515 (international or local)
707-829-0104 (fax)

We have a web page for this book, where we list errata, examples, and any additional information. You can access this page at:

http://wayneandlayne.com/bricktronics
http://oreil.ly/make-lego-arduino-projects

To comment or ask technical questions about this book, send email to:

bookquestions@oreilly.com

For more information about our publications, events, and products, see our website at *http://makermedia.com.*

Find us on Facebook: *http://www.facebook.com/makemagazine.*

Follow us on Twitter: *http://twitter.com/make.*

Watch us on YouTube: *http://www.youtube.com/makemagazine.*

Acknowledgments for John

I'd like to thank my wife for putting up with all the Lego robots scattered around the house, and for her patience as this project took me away from the family more times than I'd like; to my kids Eileen Arden, Rosemary, and Jack, for offering unlimited enthusiasm and support; to Adam and Matthew for being great collaborators and incredibly talented engineers; to our editor Brian for taking a chance with our book; and to the original writers in my family, my dad Harold Baichtal and my grandma Marion Lillie, for inspiring me to become an author myself.

Acknowledgments for Matthew

I'd like to thank my wonderful wife for her eternal patience with a project of this undertaking. My apologizes for the mess, I promise to clean it up soon! Thanks to my brother Jeremy, for the thousands of hours we've spent together building and designing with Lego. Thanks to my parents for the buckets and buckets of bricks. Many thanks to my co-authors John and Adam for their hard work and dedication.

Acknowledgments for Adam

This project was a gas. It was both fun and expanded to fill all the available space. I'd like to thank my wife for her understanding. I hope you didn't step on a Lego. I'd like to thank John, for lugging around massive totes of bricks and conjuring both models and step-by-step drawings out of them. I'd like to thank Matthew. It's nice working with a proper hacker. I'd like to thank Dick, Wayne and all the other KiCad developers for making such an awesome open source electronic design automation suite. I'd like to thank Daft Punk, for their excellent Tron Legacy soundtrack. I'd like to thank the entire Open Source Hardware community and the Arduino team. Lastly, I'd like to thank The Lego Group for fitting all that inspiration inside their little injection-molded plastic pieces.

Project: Drawbot

Figure 1-1. *The Drawbot project consists of a small, wheeled robot that creates drawings as it rolls*

Our first project consists of a Drawbot (Figure 1-1), a small robot that rolls around with a marking pen trailing behind it, leaving a line that traces the path the Drawbot took. We'll program the robot to react to obstacles by turning away and trying another path, allowing it to keep drawing even after it hits an obstruction.

We'll tackle a very basic method of creating art. Essentially, we'll program the robot to respond to input from its touch sensor to create a series of circles, arcs, and spirals, as seen in Figure 1-2.

Figure 1-2. *Robot-created art!*

This is how it works. When the bumper is pushed the first time, each motor is assigned a random speed, from 255 (full speed forwards) to -255 (full speed backwards). There is also a random timeout assigned, between 1 and 10 seconds. If the bumper is hit, the two speeds for the motors are reversed, and slowed down a little. If the motors are already going very slowly, or when the timeout runs out, new random speeds are assigned for each motor, as well as a new random timeout. Isn't art cool?

Parts List

While not as complicated as other models, the Drawbot packs a lot of detail into a small package. Let's go over what you'll need to create the robot.

Tools & Electronics

- Arduino Uno
- Bricktronics Shield (see Figure 1-3 and the "Introducing Bricktronics" sidebar)
- Several #2, 0.25" wood screws
- 2.1 mm DC Plug (we used CP3-1000-ND from Digi-Key)
- Battery pack (We used Mouser P/N 12BH361A-6R)
- Bricktronics mounting plates (see the "Attaching the Arduino & Battery Pack" sidebar)
- Clothespin, the standard wooden kind with a spring
- Marking pen
- One 8-32, 1.5" machine screw with washer and nut

INTRODUCING BRICKTRONICS

What is Bricktronics? When the three of us began working on the book, we immediately realized we'd need something to interface between the Arduino and the Mindstorms motors and sensors. Early on we experimented with breadboarding up a small interface circuit, and figured out how to add Arduino-compatible plugs onto the ends of the Mindstorms wires, but ultimately it made more sense to build our own shield. The result is the Bricktronics Shield!

The shield gives you the ability to control up to four sensors, two Mindstorms motors, as well as additional parts like Power Functions motors. It's not intended to blow the NXT brick out of the water, but to let you use the familiar Arduino environment with the NXT motors and sensors. As we use the Bricktronics Shield throughout the book, you'll get to learn about its capabilities, and we think you'll be excited about what it can do!

If that weren't enough, we also designed the Bricktronics Motor Controller (it makes its debut in Chapter Seven) which allows you to control up to five Mindstorms motors as well as additional Power Functions motors. It's cool and powerful!

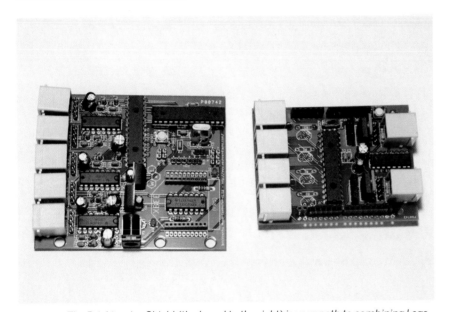

Figure 1-3. *The Bricktronics Shield (the board to the right) is your path to combining Lego and Arduino technologies! The Bricktronics Motor Controller (left) helps you control a mess of motors*

Lego Elements

We used the following Lego parts (Figure 1-4) to build our Drawbot. Gather together what you need and follow the directions later in this chapter to build your robot.

We developed the models over a few months, fixing and changing as we tested them. We're proud of them, and they work well. However, they're only one of many valid designs for each of our projects! Feel free to make adjustments and see how they change the project!

A. Touch sensor

B. 2 interactive servo motors

C. 3 Mindstorms wires (not shown in Figure 1-4)

D. 2 rims

E. 2 tires

F. 1 2M Technic beams*

G. 2 7M Technic beams

H. 4 Technic angle beams 3x5

I. 2 double angle beams 3x7

J. 2 Technic levers 3M*

K. 4 Technic levers 4M*

L. 2 Technic triangles*

M. 2 halfbeam curves 3x5*

N. 18 half bushes

O. 9 bushes

P. 5 2M cross axles (red)

Q. 3 connector pegs 3M (blue)

R. 12 cross connector pegs (blue)

S. 1 cross connector peg w/o friction tabs (beige)

T. 3 3M cross axles

U. 6 3M cross axles with knob

V. 5 4M cross axles

W. 2 4M cross axles with end stop*

X. 6 5M cross axles

Y. 1 8M cross axle*

Z. 2 tubes*

AA. 2 cross axle extensions

BB. 1 catch with cross hole

CC. 7 double cross blocks*

DD. 4 180-degree angle elements

EE. 1 Technic cross block fork

FF. 1 2x3 cross block

GG. 1 3x3 connector block*

HH. 2 3M Technic beams with snaps

II. 2 belt wheels*

* The parts marked with an asterisk are either not found in the Lego Mindstorms NXT 2.0 set or aren't included in the listed quantity. Note that the quantities found in Lego sets sometimes differ from their official numbers, due to packing errors.

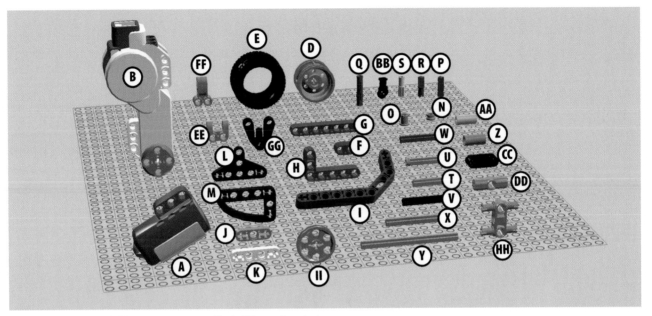

Figure 1-4. *These are the Lego parts you need to build your Drawbot*

SOURCING NON-STANDARD PARTS

Unfortunately, the Mindstorms box contains only a modest sampling of parts, especially taking Lego's vast product line into consideration. In building this model, we made an effort to keep the list of additional parts you need to buy as short as possible. If you have a small Lego collection—we're assuming a Mindstorms set at the minimum!—you can consult the list below to find the missing parts.

Note that we link to Peeron, a 3rd party site devoted to keeping track of Lego's prodigious parts catalog. Each page includes a list of merchants who offer that part. It's a great resource!

- 1 Technic 2M beam P/N 43857 *http://peeron.com/inv/parts/43857*
- 2 Technic levers 3M P/N 6632 *http://peeron.com/inv/parts/6632*
- 4 Technic levers 4M P/N 32449 *http://peeron.com/inv/parts/32449*

- 2 Technic triangles P/N 2905 *http://peeron.com/inv/parts/2905*
- 2 double cross blocks P/N 32184 *http://peeron.com/inv/parts/32184*
- 2 halfbeam curves 3x5 P/N 32250 *http://peeron.com/inv/parts/32250*
- 2 tubes P/N 62462 *http://peeron.com/inv/parts/62462*
- 1 8M cross axle P/N 3707 *http://peeron.com/inv/parts/3707*
- 1 4M Technic axle with end stop P/N 87083 http://peeron.com/inv/parts/87083
- 1 3x3 connector block P/N 111 *http://peeron.com/inv/parts/111*
- 2 belt wheels P/N 4185 *http://peeron.com/inv/parts/4185*

Alternatively, you may also choose to buy our Bricktronics Ultimate Kit that includes the above parts so you don't have to source them yourself!

Assembly Instructions

While the robot itself is relatively small, its construction can be a bit complicated. However, the following instructions will guide you through the steps.

Build the Lego Model

Let's tackle the Drawbot's Lego chassis. It's a fun little robot with three wheels and a nice big platform for holding the battery pack as well as the Arduino and Bricktronics Shield.

1. Let's begin with the bumper assembly. Connect an angle element to the triangular 3x3 block as shown in Figure 1-5.

2. Slide one of those red 2M cross axles through the block. Figure 1-6 shows where it goes.

3. Add a triangular plate to the bumper assembly, held in place by the bottom of the 2M axle, as shown in Figure 1-7.

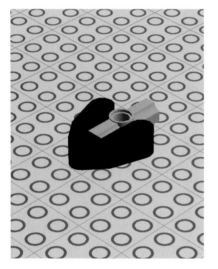

Figure 1-5. *Step 1: Bumper assembly*

Figure 1-6. *Step 2: Continuing with the bumper*

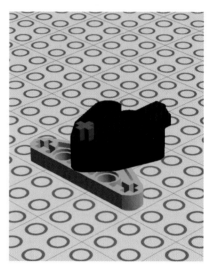

Figure 1-7. *Step 3: Adding a triangular plate*

Chapter 1

4. Add two more 2M cross axles as shown in Figure 1-8.

5. Add the angle beams shown in Figure 1-9; they'll serve as the business end of the bumper!

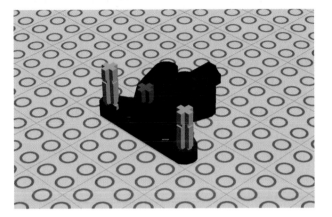

Figure 1-8. *Step 4: More axles*

Figure 1-9. *Step 5: Rounding out the bumper*

6. Add another triangular plate to the top to secure the various elements as shown in Figure 1-10.

7. Slide a 3M cross axle through the top hole of the triangular plates, leaving enough room for a half bush on the top and bottom. Figure 1-11 shows the arrangement.

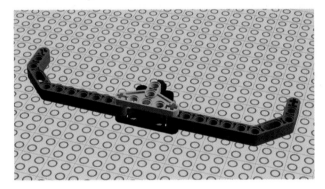

Figure 1-10. *Step 6: Topping it off*

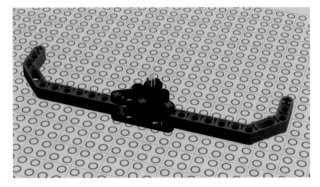

Figure 1-11. *Step 7: Inserting a cross axle with a bit coming out of both ends*

8. Add the half bushes as shown in Figure 1-12. The bumper assembly is complete!

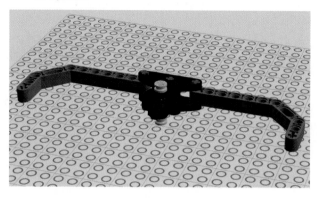

Figure 1-12. *Step 8: Finishing the bumper*

9. Next, let's work on the Drawbot's motors. Slide two 5M axles through the motor's holes, while holding an L-shaped beam in place as you do, as shown in Figure 1-13.

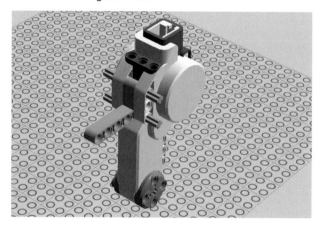

Figure 1-13. *Step 9: Attaching the axles and L beam*

10. Add two cross-axle extenders as shown in Figure 1-14.

11. Connect two 5M beams to the extenders as shown in Figure 1-15.

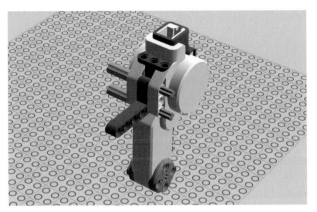

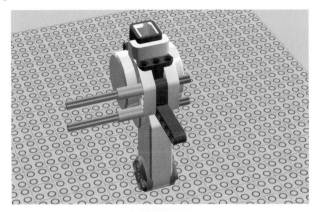

Figure 1-14. *Step 10: Adding extenders*

Figure 1-15. *Step 11: Extending the axles*

12. Next, add a 3M beam with pegs to the motor as shown in Figure 1-16.

13. Then, add a couple of tubes as shown in Figure 1-17.

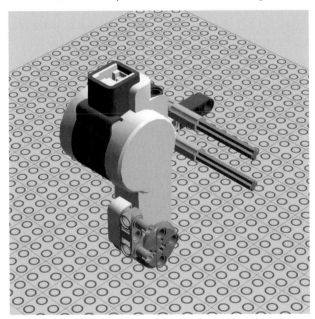

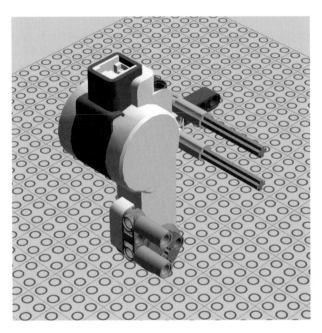

Figure 1-16. *Step 12: Attaching the 3M beam*

Figure 1-17. *Step 13: Adding a couple of tubes*

14. Then, another 3M beam with pegs! Figure 1-18 shows how it goes in.

15. Connect another motor as you see in Figure 1-19. Make sure to hold another angle beam in place as you slide the cross axles through the new motor's holes.

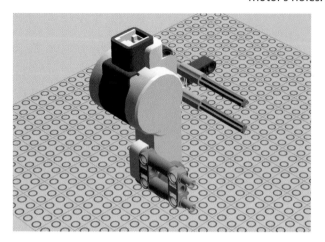

Figure 1-18. *Step 14: Attaching another beam*

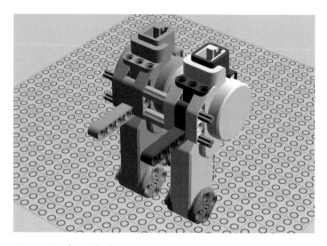

Figure 1-19. *Step 15: Connecting the second motor*

16. Secure the cross axles with bushes (see Figure 1-20). Now you're done connecting the two motors.

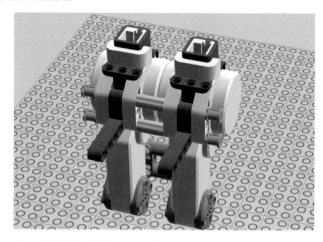

Figure 1-20. *Step 16: Securing the cross axles*

17. Next, let's tackle the rear wheel. Begin by adding a beige cross connector (the kind without friction tabs) to a catch. Figure 1-21 shows this arrangement.

18. Next, add one of the cross blocks and secure it with a 2M cross axle, as shown in Figure 1-22.

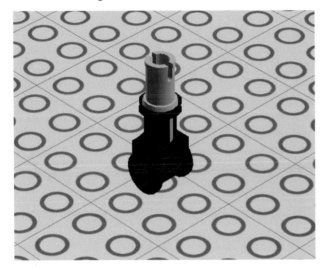

Figure 1-21. *Step 17: Starting the rear wheel*

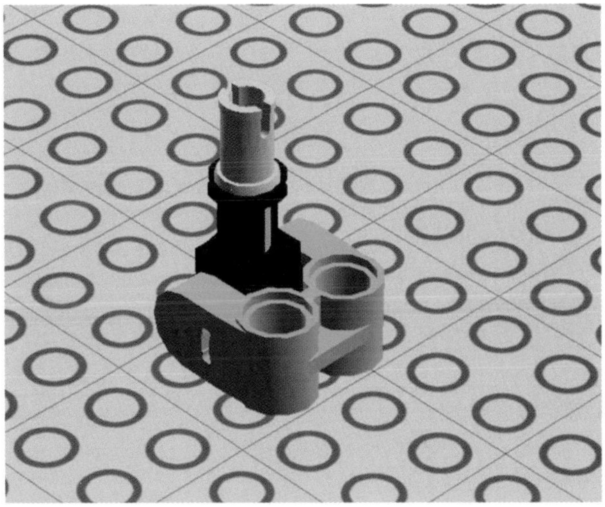

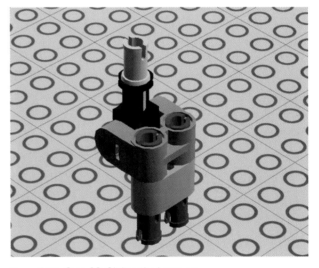

Figure 1-22. *Step 18: Securing the assembly*

19. Add two 3M connector pegs (see Figure 1-23).

20. Slide a 2M beam onto the pegs as shown in Figure 1-24.

Figure 1-23. *Step 19: Adding connector pegs*

Figure 1-24. *Step 20: Sliding the beam on*

21. Connect a 2x3 cross block to the ends of the pegs. Figure 1-25 shows how they go together.

22. Add a 2M cross axle and a belt wheel (Figure 1-26).

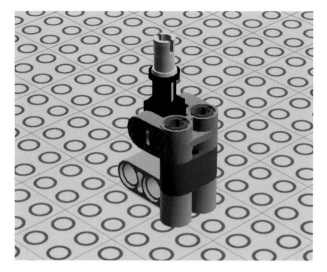

Figure 1-25. *Step 21: Connecting the cross block*

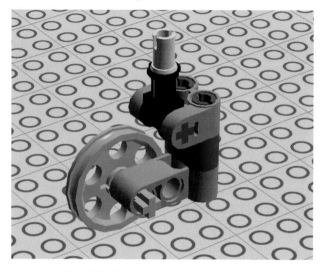

Figure 1-26. *Step 22: Attaching the wheel*

23. Connect another belt wheel (Figure 1-27). The rear wheel assembly is complete! Set it aside for now; we'll attach it at the end.

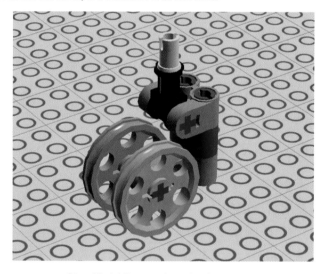

Figure 1-27. *Step 23: Adding another wheel*

Chapter 1

24. Next, let's work on the main chassis. Connect a 7M beam, a halfbeam curve, and a 4M cross axle as shown in Figure 1-28. Note that this won't stay on—you'll have to hold it in place with your finger for now.

25. Add a cross block and a 3M liftarm on the back, as shown in Figure 1-29.

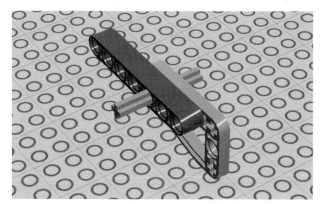

Figure 1-28. *Step 24: Beginning the main chassis*

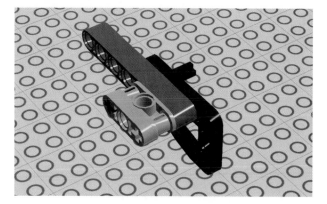

Figure 1-29. *Step 25: Attaching cross block and liftarm*

26. And another 4M axle (Figure 1-30)!

27. Connect a 3M peg to a 180-degree angle element, as shown in Figure 1-31.

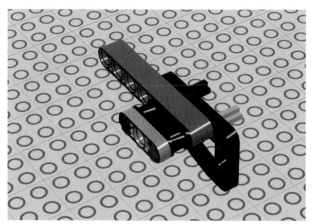

Figure 1-30. *Step 26: Adding one more axle*

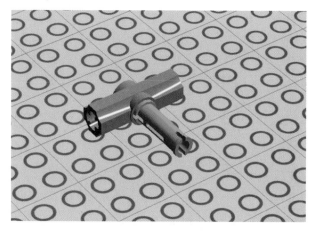

Figure 1-31. *Step 27: Connecting the peg to the angle element*

28. Add two more angle elements to the peg (Figure 1-32).

29. Next attach the angle elements to the ends of the two 4M axles as shown in Figure 1-33.

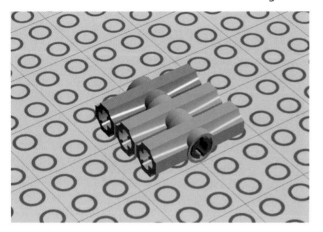

Figure 1-32. *Step 28: Two more angle elements*

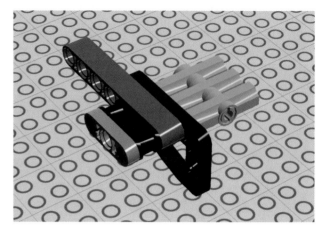

Figure 1-33. *Step 29: Connecting the assemblies*

30. Shove two more 4M cross axles into the angle elements (Figure 1-34).

31. Then, shove another halfbeam curve onto the ends of the axles as shown in Figure 1-35.

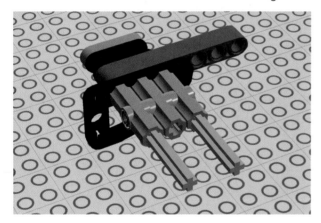

Figure 1-34. *Step 30: Two more axles*

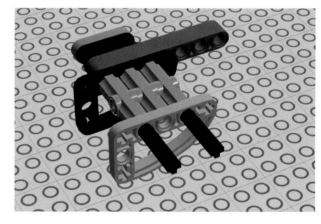

Figure 1-35. *Step 31: Adding another halfbeam curve*

32. Add another 7M beam as shown in Figure 1-36.

33. Add a cross block and 3M liftarm to the cross axles (Figure 1-37).

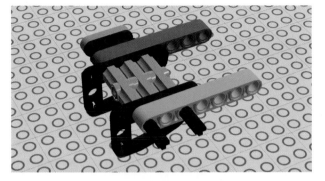

Figure 1-36. *Step 32: Adding another 7M beam*

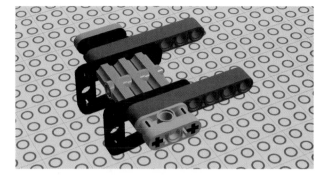

Figure 1-37. *Step 33: Adding cross block and liftarm*

34. Slide a pair of 3M axles through the holes of a cross block. Figure 1-38 shows this step.

35. Next, pop on a pair of 3×5 angle beams (Figure 1-39).

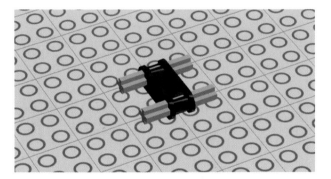

Figure 1-38. *Step 34: Connecting two axles to a cross block*

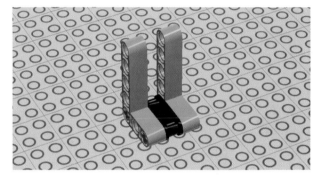

Figure 1-39. *Step 35: Attaching L-shaped beams*

36. Position the assembly you created as you see in Figure 1-40; the next step shows you how to secure it.

37. Slide a 5M axle through the cross-holes of the halfbeam curves as shown in Figure 1-41.

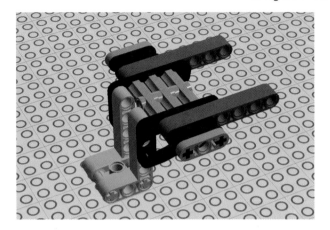

Figure 1-40. *Step 36: Lining the beams up with the assembly*

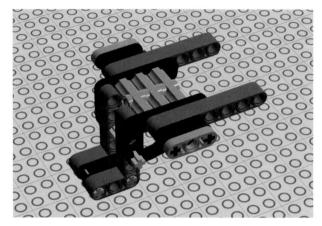

Figure 1-41. *Step 37: Securing things in place*

38. Add another 5M axle, but this time, hold a bush in place in the middle with your fingers as you slide the axle through, as shown in Figure 1-42.

39. Add half bushes to keep everything in place (Figure 1-43). The back assembly is done!

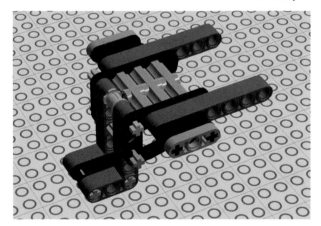

Figure 1-42. *Step 38: Adding another axle*

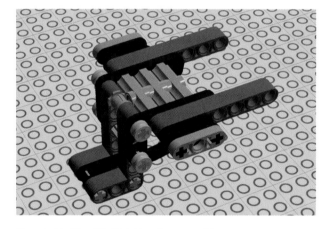

Figure 1-43. *Step 39: Finishing the assembly*

40. This is a tricky step. Hold a touch sensor in place and slide an 8M cross axle through it as well as two bushes and two half bushes. The sensor will swing freely (Figure 1-44).

41. Then secure the ends with two more bushes as shown in Figure 1-45.

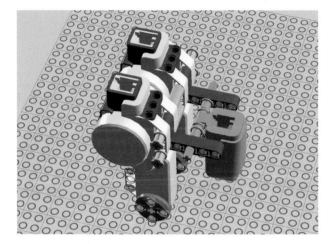

Figure 1-44. *Step 40: Connecting the sensor*

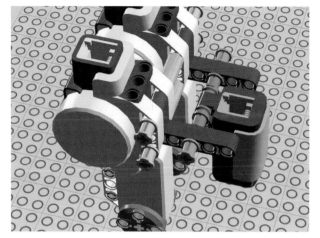

Figure 1-45. *Step 41: Securing the assembly*

42. Swing the bump sensor forward and shove in a 5M axle.

43. Add rims! Secure them in place with the 4M axles with end stops (Figure 1-47).

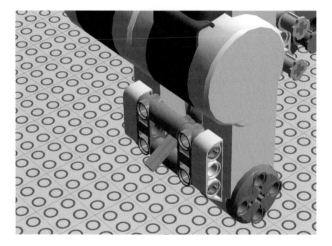

Figure 1-47. *Step 42: Adding the 5M axle*

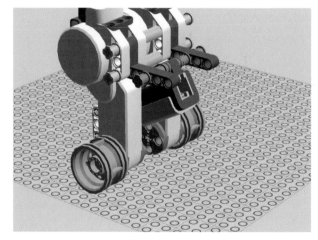

Figure 1-46. *Step 43: Adding rims*

44. Next, add tires as shown in Figure 1-48.

45. Add a couple of 4M plates, secured with 3M cross axles with knobs (Figure 1-49).

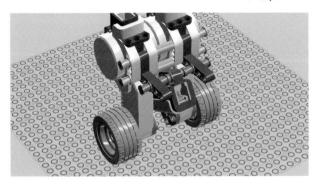

Figure 1-48. *Step 44: Then, tires*

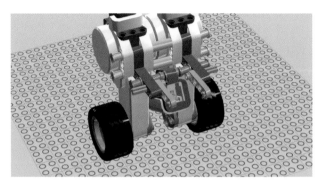

Figure 1-49. *Step 45: Adding and securing the plates*

46. Add the back assembly as shown in Figure 1-50. Note that you'll have to hold it in place until the next step!

47. Add four more 3M cross axles with knobs, as well as two cross connectors (Figure 1-51).

Figure 1-50. *Step 46: Adding the back assembly*

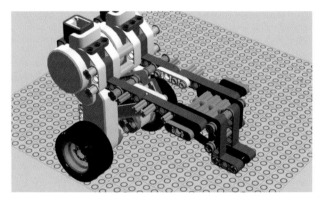

Figure 1-51. *Step 47: Adding more axles*

Chapter 1

48. Add double cross blocks as shown in Figure 1-52.

49. Add two 4M liftarms to the ends of the cross axles as shown in Figure 1-53.

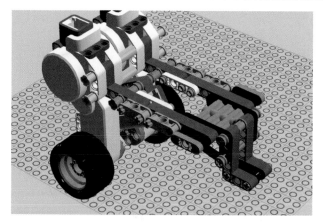

Figure 1-52. *Step 48: Adding the double cross blocks*

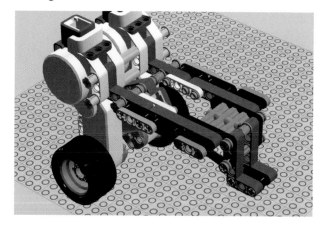

Figure 1-53. *Step 49: Adding liftarms*

50. Pop in 10 cross connectors as shown in Figure 1-54. Your Bricktronics mounting plates will connect to these pegs.

51. Add the bumper (Figure 1-55). You're almost done!

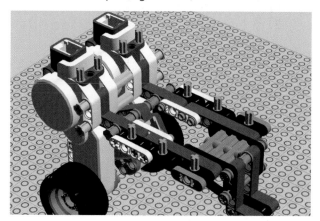

Figure 1-54. *Step 50: Adding cross connectors*

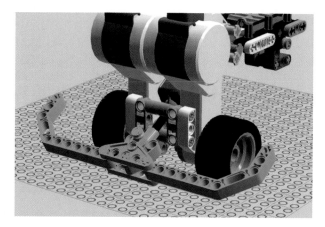

Figure 1-55. *Step 51: Connecting the bumper*

52. Connect the rear wheel as shown in Figure 1-56.

53. You're all done (Figure 1-57). Let's add electronics and the pen! Note that the robot will want to flex a bit in the middle, but the Bricktronics mounting plate will keep it rigid.

Figure 1-56. *Step 52: Connecting the rear wheel*

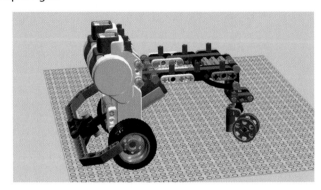

Figure 1-57. *Step 53: Admire your work!*

Attach the Arduino and Bricktronics Shield

Adding the Arduino and shield are ridiculously easy. Just connect up your Mindstorms wires to the shield as seen in Figure 1-58. Don't have a Bricktronics Shield? Don't worry, we show you how to wire up a breadboarded equivalent in Chapter Ten, Advanced Techniques.

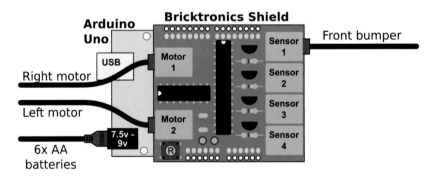

Figure 1-58. *Connect the Mindstorms wires as you see here*

Chapter 1

ATTACHING THE ARDUINO & BATTERY PACK

In order to attach the battery pack (and therefore the Arduino and Bricktronics Shield), you'll need to use a connector plate. There are a number of these floating around on the Internet but we designed our own (Figure 1-59). They consist of thin pieces of wood or acrylic, drilled with Technic-compatible holes as well as screw-holes for connecting the plate to battery packs and Arduinos.

If your battery packs didn't come with DC plugs, attach one like the one in the parts list. Remember to solder the positive wire of the battery pack to the central terminal, and the negative wire to the outer terminal.

The first step is to screw the battery pack to the plate using the quarter-inch screws we specified. Next, find some

available Technic holes that line up with the plate and thread the robot's cross connectors through the holes in the plate. Secure the cross connectors with half bushes.

Next, repeat the above steps with the Arduino. In the Drawbot model, the Arduino plate attaches to the four free cross connectors plugged into the motors (Figure 1-61). You're set!

If you'd like to create your own connector plates, download the design (*http://www.wayneandlayne.com/bricktronics/ mounting_plate*) and laser cut your own, or alternatively, you can buy the plates as part of our Bricktronics Ultimate Kit.

Figure 1-59. *These connector plates, prototyped in wood, allow you to connect your electronics to Lego beams*

Figure 1-60. *Attach the battery pack to the robot's platform; note that this photo shows a prototype cut out of thick wood, so the bushes aren't necessary*

Figure 1-61. *The Arduino's mounting plate fits on top of the battery pack and is secured by four cross connectors*

Attach the Pen

Next, let's connect the marking pen to the robot with the help of a clothespin. Take the machine screw we specified and attach the clothespin to one of the battery pack mounting plate's free Technic holes. Depending on the clothespin you used, you may be able to thread the screw through the hole in the spring; otherwise, position the screw next to the spring as seen in Figure 1-62. This arrangement will give the pen the maximum flexibility to move as the robot changes direction.

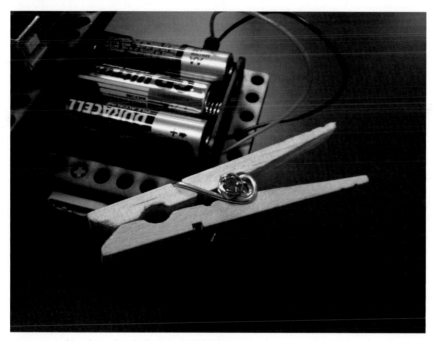

Figure 1-62. *Attaching the clothespin to hold the pen*

Program the Robot

Coding up an Arduino *sketch* (as Arduino programs are called) can be surprisingly easy. You may not know how to program, or understand the rules most programming languages operate by, but getting to this point may be quicker than you think. We can't teach you how to program in this book, but we can get you to a place where you can learn for yourself, either by following tutorials or by adapting other hackers' programs for your own use.

Processing and Arduino

The language and development environment used to program the Arduino board is also called Arduino (Figure 1-63), and it's inspired by another language called Processing. Casey Reas and Benjamin Fry created Processing in 2001 as a tool to get nonprogrammers such as artists interested in programming. (Processing is based on Java, while Arduino is based on C++; this means that you won't be able to run most Processing code on Arduino or vice-versa.)

Processing features a quick turnaround between coding and results, allowing rookies the rapid gratification of seeing their Processing sketch displayed on screen after only a few seconds. As open source initiatives, Processing and Arduino encourage the sharing of code, allowing relative newbies to quickly get up to speed on how to program their boards.

For the purposes of this book, the information presented in this section is mostly informational—all you have to do is download the code for the robots in this book, though we encourage you to play around with the program. You'd be surprised what you can learn just by modifying someone else's code! We'll get into Arduino in much more detail in Chapter Three.

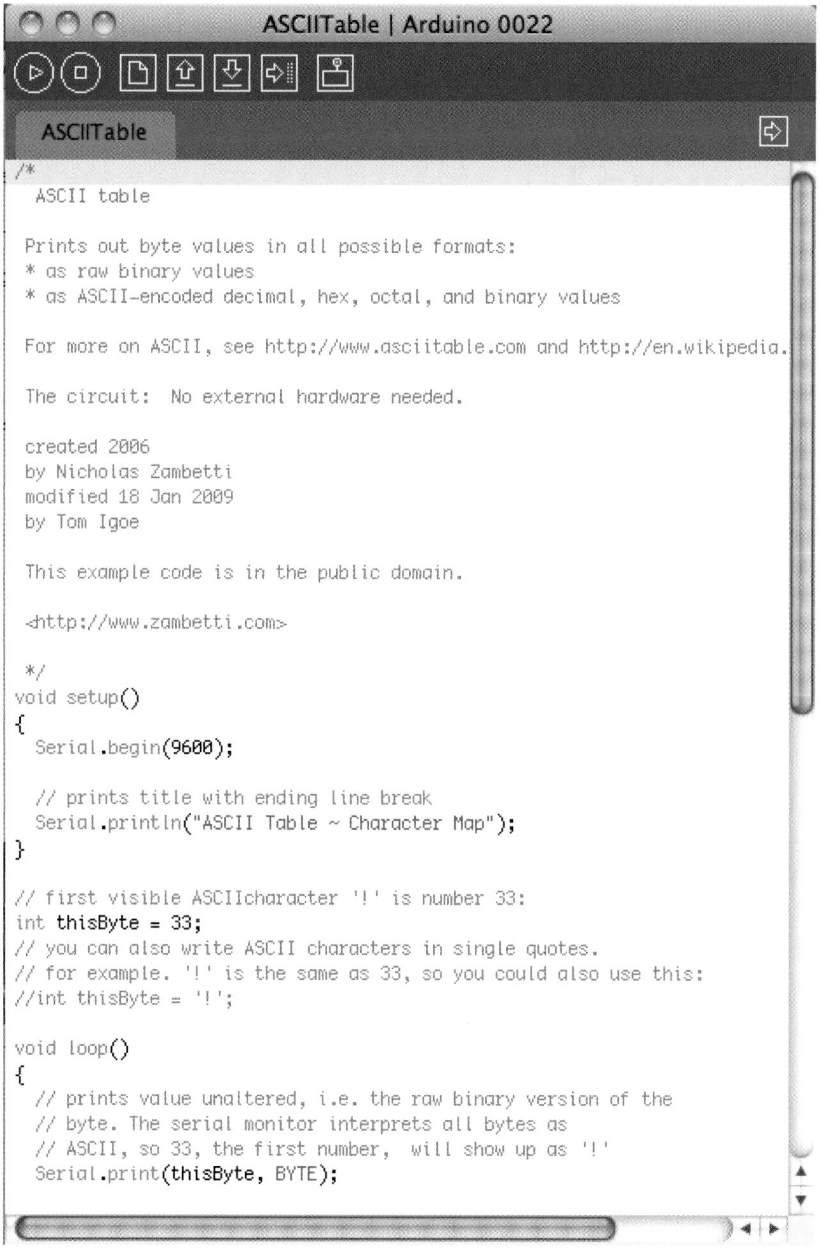
ASCIITable | Arduino 0022
ASCIITable
/*
ASCII table
Prints out byte values in all possible formats:
* as raw binary values
* as ASCII-encoded decimal, hex, octal, and binary values
For more on ASCII, see http://www.asciitable.com and http://en.wikipedia.
The circuit: No external hardware needed.
created 2006
by Nicholas Zambetti
modified 18 Jan 2009
by Tom Igoe
This example code is in the public domain.
<http://www.zambetti.com>
*/
void setup()
{
Serial.begin(9600);
// prints title with ending line break
Serial.println("ASCII Table ~ Character Map");
}
// first visible ASCIIcharacter '!' is number 33:
int thisByte = 33;
// you can also write ASCII characters in single quotes.
// for example. '!' is the same as 33, so you could also use this:
//int thisByte = '!';
void loop()
{
// prints value unaltered, i.e. the raw binary version of the
// byte. The serial monitor interprets all bytes as
// ASCII, so 33, the first number, will show up as '!'
Serial.print(thisByte, BYTE);

Figure 1-63. *Arduino code might look scary, but it's actually easy for nonprogrammers to learn*

Setting Up the Programming Environment

In order to upload code to your Arduino, you must take a few minor steps to make sure you're ready.

1) Download the Arduino Programming Environment

Your first step is to download Arduino. It's simple! Go to *http://arduino.cc/ en/Main/Software* and choose your operating system to download what you need. If you have an older machine, you can even grab legacy versions of the software, though we don't necessarily encourage this.

2) Connect the Board

Use a USB A-B cable to connect your Arduino Uno (the Leonardo uses a microUSB cable) to the computer. Note that, through the magic of USB, your board will be powered through the cable, so you won't have to worry about the power supply while you're loading code.

3) Configure Arduino

Go to the Tools menu in the menu bar and choose your board type as well as which serial port you want to use (although Arduino uses USB, it looks like a serial port—the same kind of connection used for dialup modems). If you don't know which serial port Arduino is using, unplug the Arduino from your computer, look at the menu, and make a note of which serial ports are in the list. Next, plug the Arduino back into your computer, and use the serial port that wasn't there last time you looked.

4) Load the Code

Use the File menu option to load the sketch you want. Try this out now with the Blink sketch (File | Examples | 01.Basics | Blink). Upload the sketch by clicking in the right-arrow icon in the toolbar or by choosing File | Upload. One of the Arduino's on-board LEDs will flash frantically as the code uploads, but when it's done (it should just take a few seconds unless you have a monster sketch) you'll be good to go!

Don't Forget Your Libraries!

One thing many new Arduino hackers forget is to make sure they have the right libraries for their sketch. A software library is a grouping of computer code that is intended for reuse. A library can provide a layer of abstraction, reducing the need to understand every line of code inside of it.

While there are about twenty libraries that come with the Arduino software, there are many more available online. Some of them require specific hardware, but some do not. There are libraries for connecting to LCD screens, and outputting black and white TV signals. There are motor control libraries and PS2 keyboard libraries. There are libraries for generating audio, keeping track of the date and the time, and using SD cards. There are even libraries for connecting to the Internet.

They're easy to install—simply download the library, unzip/extract it if needed, and then copy it into a *libraries* subdirectory of your "sketchbook" directory. To find this directory, go into Arduino's preferences (Arduino | Preferences on Mac, File | Preferences on Linux or Windows), and note the Sketchbook Location listed there. If there is no *libraries* subdirectory, you'll have to create it. Finally, every time you add a library you'll have to exit Arduino and restart the application. There are more detailed instructions at the Arduino website if you need more help.

Arduino libraries are easy to write—if you find yourself writing the same code in multiple programs, it may make sense for you to create a library. The extra work consists of a little standard setup code, and then some labeling of what parts of the library should be able to be accessed from other programs, and which parts should be protected. There's a good tutorial on adapting code into a library at *http://arduino.cc/en/Hacking/LibraryTutorial*.

The final step is to program the Arduino with the code needed to operate the Drawbot. First, however, you need to download and install the Bricktronics library. You can find it at *http://www.wayneandlayne.com/bricktronics*.

The Arduino Sketch

The Drawbot code is relatively simple. The main loop checks if a timeout has occurred. If it has, it randomizes the motor speeds and chooses a new random timeout. It also checks if the front bumper has been hit. If it has, it reverses the direction of both motors, slows them down a bit, and checks if the motors are going very slowly. If they are, it randomizes the motor speeds and chooses a new random timeout.

This sketch is included in the sample code that you can download for this book (see "How to Contact Us" in the Preface for more information on this).

With the Arduino connected to your computer with USB (and the battery pack disconnected), open the sketch up in Arduino, then upload the code by clicking in the right-arrow icon in the toolbar or by choosing File | Upload. Unplug the Arduino from the computer, plug in the battery pack, set down the robot, press the front bumper, and watch your robot go!

```
#include <Wire.h> ❶
#include <Adafruit_MCP23017.h>
#include <Bricktronics.h>

// Make: Lego and Arduino Projects
// Chapter 1: Drawbot
// Website: http://www.wayneandlayne.com/bricktronics/

long timeout = 0;

Bricktronics brick = Bricktronics(); ❷
Motor r = Motor(&brick, 1); ❸
Motor l = Motor(&brick, 2);
Button front = Button(&brick, 1); ❹

void setup() ❺
{
  randomSeed(analogRead(A3)); ❻
  brick.begin();
  r.begin();
  l.begin();
  front.begin();
}

void bumpers_hit() ❼
{
 int r_speed = r.get_speed() * -0.9;
 int l_speed = l.get_speed() * -0.9;
 r.set_speed(r_speed);
 l.set_speed(l_speed);

 delay(500);
 if (abs(l_speed) < 25 && abs(r_speed) < 25)
 {
   timeout_expired();
 }
}

void timeout_expired() ❽
{
  r.set_speed(random(-127, 127));
  l.set_speed(random(-127, 127));
  timeout = millis() + random(1, 10) * 1000;
}

void loop() ❾
{
 if (timeout != 0 && timeout < millis()) ❿
 {
   timeout_expired();
 }
```

```
  if (front.is_pressed()) ⓫
  {
    bumpers_hit();
  }
}
```

❶ These 3 lines let the Arduino sketch use the Bricktronics library code that simplifies working with motors and sensors.

❷ The Bricktronics object manages many of the Bricktronics library calls.

❸ The `r` and `l` Motor objects correspond to the motors plugged into Motor Port 1 and 2, respectively, of the Bricktronics Shield.

❹ The front Button object corresponds to the button plugged into Sensor Port 1.

❺ The `setup()` function is called once, on power on. Here, it's used to initialize all the Bricktronics objects.

❻ This line *seeds* the pseudorandom number generator with an analog reading from a disconnected pin, A3. This helps the numbers be more random.

❼ This function is called when the bumpers are hit. The idea is to reverse each motor, and slow it down a bit, but if it slows down too slowly, to start over with new random values, so it doesn't get boring.

❽ This function sets the motors going at a random speed, and it determines a timeout between 1 and 10 seconds, after which it will be called again.

❾ In Arduino, your `loop()` function is called over and over again, until the Arduino runs out of power.

❿ If the timeout has been set, and the current time is past the timeout, run the `timeout_expired()` function.

⓫ If the front bumper is pressed, then call the `bumpers_hit()` function, which reverses the robot unless the motors are going too slowly.

The Next Chapter

In Chapter Two we'll brush up on our Lego lore. What's up with the Mindstorms set? When was it introduced? What do you get and what can you do with it? The answers to these questions (and more!) will bring you up to speed on this fascinating set.

Anatomy of Lego Robotics

2

Before we delve any deeper into the specifics of how Arduinos interact with Mindstorms, let's check out the particulars of Lego's robotics technology.

While the Lego Group has produced many robotics sets over the years, their Mindstorms NXT 2.0 product represents their biggest, most popular, and most sophisticated offering to date, and so we're focusing on that set in this book.

It's really amazing what has been done with Mindstorms, with creations ranging from knitting machines to constructs that can sort bricks by color and shape to recreations of legendary computing engines from the 19th century. Even so, when you consider that it's a construction set coupled with a full-fledged robotic prototyping system, you begin to understand why it's so popular.

Figure 2-1. *Hans Andersson's Time Twister shows the potential of Lego robotics. Credit: Hans Andersson*

Hans Andersson's Time Twister clock (*http://tiltedtwister.com/robots.html*, Figure 2-1) serves to demonstrate the capabilities of this robust system. Using only two NXT bricks and five motors, Andersson controls a series of interlocked layers that can be twisted to create the digits of a clock. It's a virtuoso robot with an elegant design that maximizes capabilities with a bare minimum of resources.

So, what's Lego robotics all about? Let's jump in and find out!

Mindstorms

The Lego Group turned to academia to help them create the ultimate kids' robotic construction set, and they did it right—they engaged the legendary MIT Media Lab in a multi-decade collaboration, beginning in 1985. In the early '90s, Lego opened an office in Cambridge to facilitate the partnership. The first prototype "Programmable Brick" was developed in 1998, fulfilling the goal to create a Lego brick with a microcontroller inside.

The Mindstorms Robotics Invention System (RIS) product was released shortly thereafter, and featured the RCX (Robotic Command eXplorer) microcontroller brick, a yellow plastic module with Lego studs on the top, four buttons, and an LCD screen. The Lego Mindstorms NXT set, released in 2006, took the best ideas of the RIS and fixed many of the flaws, and 2009's NXT 2.0 set (Figure 2-2) put the finishing touches on what has become Lego's most successful project.

Figure 2-2. *The Mindstorms NXT 2.0 set represents the culmination of a 20-year quest to build the ultimate robotics set*

The next sections look at what you get in a Lego Mindstorms NXT set.

Chapter 2

The NXT Brick

The microcontroller accompanying the Mindstorms set is called the NXT Intelligent Brick. As seen in Figure 2-3, it's a plastic box with a 100x64 pixel LCD screen, a speaker, four buttons, three motor ports, and four sensor ports, as well as a USB port for programming the unit. Its 32-bit ARM7 microprocessor, flash memory, Bluetooth support, and built-in battery pack make it ideal for prototyping robots.

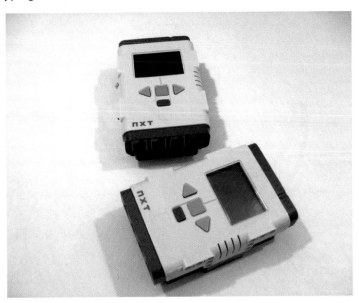

Figure 2-3. *Don't mistake the NXT brick for a toy—it's a very robust robotic control system*

Despite the sweetness of the NXT brick, however, we have chosen to focus on Arduino as the brains of the robots described in the book. Therefore, we won't be hacking the NXT bricks much, but rest assured the NXT is a potent tool for developing your next project.

NXT VS. RCX

Though obsolete since the release of the first NXT set, the RCX brick (Figure 2-4) has seen continual use since then. Simply put, it was an amazing product in 1998 and remains solid. Putting it in perspective, teens' robotics championship FIRST Lego League used to have a separate category for RCX models, to go easy on kids with older sets, but the category was eliminated when it was discovered that RCX models competed quite well with NXT bots. They have a slightly slower processor, no USB interface or Bluetooth, but still control motors and receive input from sensors. Despite the age of these bricks—six years as we write this—you'll see many RCX bricks floating around. Don't be afraid to use one!

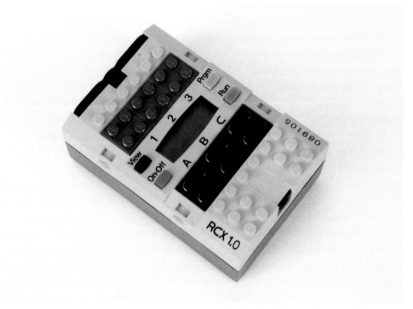

Figure 2-4. *The original RCX brick holds up well against its more recent brethren. Credit: Pat Arneson*

Sensors

The standard Mindstorms NXT 2.0 set comes with four sensors (Figure 2-5): two touch sensors, a color sensor, and an ultrasonic sensor.

The light sensor uses an LED for illumination or to reflect light back to a color sensing component, allowing it to identify the color of the item scanned.

The ultrasonic sensor consists of a sound emitter and a microphone. The emitter emits pulses of human-inaudible sound, and the microphone picks them up and determines whether an object exists within the range of the sensor.

Finally, the touch sensor is essentially a button.

Figure 2-5. *The color sensor, touch sensors, and ultrasonic sensor are found in the NXT 2.0 set*

One of the intriguing aspects of these sensors is being able to use them for more than just their obvious applications. For instance, you can use a touch sensor as a button, and the color sensor as a proximity sensor. Even a Lego motor can also be used as a sensor. It sends feedback to the NXT brick telling it how far the hub of the motor is turned, allowing it to function as a rotation sensor. These are just a few of the many different tricks that Mindstorms hackers employ with these sensors.

Motors

Each NXT set comes with 3 motors, called Interactive Servo Motors (Figure 2-6) in Lego parlance—but don't get them confused with hobby servos! They're relatively powerful motors with optical encoder feedback, allowing the position of the hub to be determined to within one degree.

When interfacing motors with Arduino projects, many people use DC motors, hobby servos, or stepper motors, depending upon what they need. DC motors are great for providing motion without precision, and are often used in simple toys. Hobby servos are small, geared motors with feedback. They usually have a limited range of motion, but they can be precisely directed to a specific angle. Stepper motors move in precise steps, and can be directed step by step. They are often used in CNC and 3D printing applications. The motors in an NXT are a little different than any of those. They're DC motors, but they have feedback so an Arduino can get some information about how the motor is moving.

Figure 2-6. *You'll get three motors in the Mindstorms set, but don't let that limit you!*

The motors' internals (Figure 2-7) offer a lot of fascinating detail. There's a DC motor that has been geared down with a 48:1 ratio, optical quadrature encoders, and a printed circuit board (PCB). The NXT motor connector has six wires. Two of them are used to drive the motor. When 9V is applied between the two inputs to the motor, the orange hub rotates in one direction. However, when 9V is applied between the two inputs the other way, changing the polarity, the hub rotates in the opposite direction. Lowering the voltage or using pulse-width modulation (PWM) on the two motor wires will cause the orange hub to rotate slower. See the "About the L293D Chip" sidebar in Chapter Four.

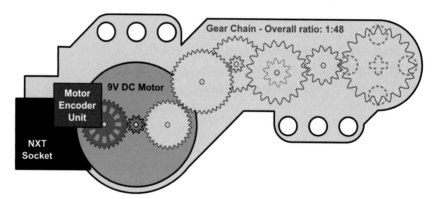

Figure 2-7. *More than merely a motor, the Mindstorms servo's speed and position direction can be monitored*

Chapter 2

The DC motor is also connected to an incremental optical quadrature encoder. This device provides information on the speed and direction of the DC motor. It consists of a circle of black plastic with slots cut in it. There is an optical fork that can tell how fast the black circle rotates due to the optical beam being repeatedly blocked and unblocked. Because there are two sensors inside the fork, it can also tell which direction the circle is rotating.

The optical sensors are connected to a PCB with some components on it. The PCB contains some circuitry that cleans up the quadrature encoder signals, which are powered by two lines on the connector: ground and power. While the power input to the motor connector is nominally 4.3 volts, we've investigated this and it appears that the circuitry is fine with 5 volts.

These clean signals are provided as two outputs on the NXT motor connector. Each line has two states, either "optical beam blocked" or "optical beam not blocked." A microcontroller can watch these changing states and keep track of the speed and direction of the motor. This feedback allows us to precisely control the motor. Without it, we cannot tell if the motor is stalled, or if the attached mechanism is being moved by external forces, like gravity, or your fingers.

Wires

Mindstorms wires are essentially proprietary RJ12 connectors with off-center tabs. Lego includes several wires of various lengths with the set. They're reversible and can be used with either motors or sensors. In Chapter Ten, Advanced Techniques, we'll explore wires in a little more depth and demonstrate how to crimp a non-proprietary plug onto the end.

Technic Beams, Mechanics & Connectors

If you look at the details of the RIS set on Peeron (http://www.peeron.com/inv/sets/9719-1) you may notice the assortment of "System" parts in the inventory—these are bricks with the classic stud-and-tube connection method. You won't find any studs in the NXT set. Instead, Lego employs a more robot-friendly system called Technic.

Recognizing the fragility of System models, the Lego Group began developing a more machinelike Lego experience in 1977, and by 1984 the line was known as Technic. It consists of Lego bricks pierced with holes, as well as studless beams, gears, axles, and other machine parts. Technic models offer increased durability and playability, and hence make better robots.

The following Technic elements may be found in the NXT set:

Beams

Plastic girders pierced with Technic holes, as seen in Figure 2-8. They typically come in odd-numbered sizes. The 7M beam, for instance, has seven Technic holes arranged evenly along its length. (See the sidebar "Measuring Lego" for more information on how Lego elements are measured.) Many are straight, but some, called liftarms, are beams that are bent in the middle. Technic beams are the default way to build robots in Lego.

Figure 2-8. *Technic beams represent the basic building blocks of most Lego robots*

Pegs

> The set provides a variety of pegs for connecting beams together. They come in a variety of lengths and offer different options for friction—some allow rotation, others do not.

Cross axles

> You'll find over 40 of these in an NXT set. They're used as actual axles for wheeled robots, but also serve as physical supports for larger models.

Connectors

> Axles are secured with bushes and other elements, which have a variety of configurations including Technic holes and pins, angles, friction options, and so on.

Gears

> Lego offers a robust assortment of gears for transmitting rotational energy from the motors' hubs to wherever it's needed in the model. While the gears included with Mindstorms set will get you through most projects, you may find yourself pining for some of the neat options like the 40-tooth gear we use in Chapter Four.

Wheels and treads

> A robot needs a way to get around, doesn't it? The NXT set includes four wheels, four tires, and two rubber tank treads.

The parts described above don't represent an exhaustive inventory of the NXT 2.0 set, but they're the primary tools you'll be using when building robots.

MEASURING LEGO

As one might gather by reading this book, Lego bricks are precisely engineered and have matured into an elegant and cohesive building system. As such, you'll find a great deal of consistency in the measurements of the various elements.

As seen in Figure 2-9, the distance between the centerpoints of two studs is 8.0mm. The height of a normal brick is 9.6mm. The gap between bricks when connected side by side is 0.2mm, or 0.1mm on each side. The stud diameter is 4.8mm, and a stud is 1.8mm tall. The Technic holes are barely larger than the stud diameter of 4.8mm, so they can be snapped together, and a cross axle is barely thinner than the stud diameter of 4.8mm, so it can move freely inside holes. When a Technic beam has multiple holes along the side, their centers are spaced 8.0mm apart, the same as the studs on top. Because many of these numbers are units of 0.8mm, bricks, axles, and beams can be connected in multiple ways.

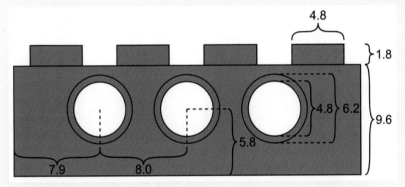

Figure 2-9. *Lego bricks' measurements are consistent across the company's entire product line*

Oftentimes you'll encounter a number like 7M in reference to a Lego brick. This refers to the length of the element in standard Lego widths. For instance, the length of a 4M cross axle equals the width of 4 beams lined up side-to-side. The 5M Technic beam sports 5 holes along its length, and conforms to the widths of 5 beams next to each other.

The next time you're playing with Lego bricks, take the time to appreciate all the clever ways that they lock together. It's a great system—use it to your advantage!

Expanding on the Mindstorms Set

While robust, the Mindstorms box doesn't hold everything. With the diversity of bricks available, you simply must avail yourself of other options to find the full potential of the set. Unfortunately, Lego only offers the one Mindstorms set, not counting educational versions with slightly different contents. The good news is that some of Lego's other products can be integrated into a Mindstorms robot, and even better, third-party companies have created their own electronics and bricks that add more capabilities. In the following section we cover some of the numerous options for adding to your Mindstorms arsenal.

Buying More

Throughout this book, we're featuring some pretty intricate models composed of more bricks than some readers might own. Here's where you can purchase some more parts:

Pick a Brick (PaB)

Lego sells individual bricks through their online and bricks-and-mortar Pick a Brick stores (Figure 2-10). While they don't have everything, the color choices and diversity of parts are unmatched anywhere. If you have a PaB store near you, be sure to check it out.

Figure 2-10. *The Pick-a-Brick wall at a Lego store offers more bricks in more colors than anywhere else. Credit: Max Braun*

Lego Education

An oft-overlooked resource, Lego Education (*http://www.legoeducation. us*) sells to STEM teachers and FIRST groups, and offers numerous Lego sets and add-ons like the pneumatic set, not available anywhere else. They even offer third-party products that are compatible with the core set.

BrickLink

The biggest online Lego brick trading site is BrickLink (*http://www. bricklink.com/*). Need an old part that's not for sale any more? Check BrickLink first. You can search sales by part number or color, and can even post a request to buy a specific brick. The eBay of Lego, BrickLink is the best way to buy individual parts.

Chapter 2

Buying sets

Our favorite way of finding the parts we need. This is how you do it. First, check the set inventories at *http://peeron.com/* and find a current set offering the parts you want. Buy, build, then break apart!

Garage sales

You can't plan on finding a treasure trove of bricks—much less specific parts—but if you can, be sure to check out garage sales offering Lego parts. Chances are you'll get entire boxes of bricks for a song, and who knows what rare or discontinued parts you'll find?

Add-On Electronics

Fans of the Lego Group acknowledge that the company can't provide *everything* that *every* fan wants. They turn to third party companies who create and sell everything from molded bricks to decals to electronic accessories. The latter category is particularly flush with cool Mindstorms sensors because electronics isn't Lego's core competency and they're happy to provide only the basics, leaving the advanced stuff to these independent companies to develop and sell.

Here are some examples:

Mindsensors Motor Multiplexer

Mindsensors (*http://www.mindsensors.com/*) offers a lot of great products including this board, which allows you to control four Interactive Servo Motors with one NXT port.

Dexter dCompass

Dexter Industries' Three-axis digital compass (*http://dexterindustries.com/dCompass.html*) allows your robot to always know where magnetic north lies.

HiTechnic Acceleration / Tilt Sensor

HiTechnic sells a sensor (*http://www.hitechnic.com/cgi-bin/commerce.cgi? preadd=action&key=NAC1040*) that tells you how fast your robot is moving and whether it's tilting.

Dexter Pneumatic Pressure Sensor

Combine this sensor (*http://dexterindustries.com/Products-dPressure.html*, Figure Figure 2-11) with Lego's Pneumatics Add-On Set to monitor the amount of pressure in your system.

Figure 2-11. *Dexter Industries' pressure sensor measures up to 70 psi of pressure. Credit: Dexter Industries*

Vernier Environmental Science Package

This educational set (*http://www.vernier.com/products/packages/engineering-nxt/environmental/*) includes temperature probes, a pH sensor, a soil moisture sensor, and other environment-related add-ons.

Dexter WiFi Sensor

Connect your robot to the Internet with Dexter Industries' WiFi Sensor (*http://dexterindustries.com/wifi.html*)! Possible applications include having your bot send out tweets or to allow others to control the robot's activities from afar.

There are dozens of alternatives out there—don't be shy about checking them out because they'll make your robot all the sweeter!

Third-Party Bricks

To a lesser extent, you'll also find companies and individuals selling bricks in shapes, colors, and materials not found in Lego's vaults, catering to fans who want to add a unique twist to their models. Here are just a handful of examples:

- The Brick Machine Shop (*http://www.bricklink.com/store.asp?p=Eezo*) sells stainless steel duplicates of Technic beams, gears, and axles.

- Chrome Block City (*http://www.brickshelf.com/cgi-bin/gallery.cgi?f=415736*) takes standard Lego parts and chromes them.

- Fans of 3D printing site Thingiverse often upload Lego-compatible creations like Philipp Tiefenbacher's Parametrized Lego Bricks (*http://www. thingiverse.com/thing:591*), as well as Brian Jepson's MyBeam: Technic-Compatible Objects (*http://www.thingiverse.com/thing:29989*).

There are many such explorations that can be found on the Internet.

Non-Mindstorms Lego Bricks

You don't have to leave the cozy realm of official Lego products to add capabilities to your Mindstorms projects. As Technic found success, its components have made their way into numerous sets.

Bionicle/Hero Factory

These "action figure" style models (Figure 2-12) seem simplistic but pack dozens of unique, Technic-compatible bricks that can be added very easily to your model. If you're looking to get away from the classic gray beam look of Technic, this is a sure bet.

Figure 2-12. *See Bionicle for what it really is: a ready supply of unique Technic-compatible bricks*

System

You'll also find lots of beams and pegs in System models, especially vehicle models intended to be handled and played with. Many also include studded Technic bricks, merging the two formats into one and allowing you to show some System love to your Mindstorms robot.

Power Functions

Interestingly, while Mindstorms focuses on autonomous robots, its Power Functions product line (Figure 2-13) lacks a microcontroller and instead focuses on switch-actuated and remote-controlled robots. It's the default automation package for high-end Technic models like trucks with working cranes. Unfortunately, its components cannot be natively plugged into Mindstorms ports and vice versa. But don't let that stop you! There are several wonderful Power Functions components and later on in the book we'll show you how to integrate them with Mindstorms.

Figure 2-13. *A parallel development to Mindstorms, Lego's Power Functions line helps you create cool motorized effects in your models*

Battery Pack

A standalone battery pack with built-in power switch. As Mindstorms doesn't have a battery pack separate from the NXT brick, this pack is required for any Power Functions project, unless you find another way of powering your robot.

DC Motors

Power Functions motors lack the position- and speed-control mechanisms built into the Mindstorms Interactive Servo Motors. This can be a convenience if you don't need this feature! Sometimes you just need to turn an axle and Power Functions' motors give you that with a smaller form factor than the servos.

Chapter 2

Remote Control

Another component that you won't find in the Mindstorms box, these simple remote controls offer a very stripped down control of your robot by streaming commands via infrared, much the way a TV remote control works.

IR Receiver

This module gets built into your robot to accept signals from the remote control.

Linear Actuator

The Power Functions linear actuator converts rotational motion to linear, letting you push or pull items using a motor.

Pneumatics

Finally, there is Lego's pneumatics set (http://www.legoeducation.us/eng/ product/pneumatics_add_on_set/1572). Not really part of Power Functions, it instead serves as its own product. The pneumatics set uses pressurized air tanks and hoses to move the parts of your robot.

Programming Your Brick

The default way to program an NXT brick involves NXT-G, a kid-friendly programming environment that uses Lego-like blocks (Figure 2-14) to organize commands. Say you want to run a motor for ten seconds. You drag and drop a Move block from the palette onto your work area and customize its settings. For instance, you could set the Move block to make the motor spin counterclockwise with 75% power for 10 rotations. The advantage of this graphical interface is that you can immediately start programming without fiddling with a programming language's syntax.

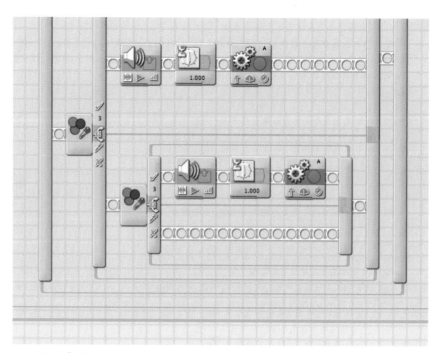

Figure 2-14. *Rather than employing the text-based interface of most programming languages, NXT-G uses programming blocks*

While newbies may appreciate NXT-G, many experienced hackers sought out a more traditional programming environment by uploading nonstandard firmware onto the NXT brick, ranging from C to Python to Java. The following list includes some of the more popular alternatives but it is by no means exhaustive.

NXC

One of the most popular firmware options, NXC (Not eXactly C, *http://bricxcc.sourceforge.net/nbc/*) uses syntax based on the ubiquitous programming language C, giving experienced programmers more precise control of their robots. (Incidentally, the Arduino is also programmed using a variant of C.)

RobotC

Another C variant, RobotC (*http://www.robotc.net/*) has been adapted for multiple robotics platforms including VEX and NXT, and boasts the ability to port the code from one platform to another. RobotC is proprietary commercial software.

leJOS

This firmware includes a Java virtual machine and is sometimes used to teach Java to beginning programming students. leJOS (*http://lejos.sourceforge.net*) is also open source, giving users the option to contribute to the platform's development.

ICON

Firmware that allows users to program the NXT brick directly from the brick's button interface, without the need for a computer. ICON (*http://www.teamhassenplug.org/NXT/ICON/*) was developed by Team Hassenplug, one of the foremost groups of Mindstorms hackers with numerous impressive robots.

The Next Chapter

With two projects under your belt, as well as a grounding in Lego lore, it's time to delve into the Arduino world in Chapter Three. We'll learn about the history of the Arduino project as well as the constellation of add-on boards and variants that have been created.

Arduino Interlude

In the first two projects you used an Arduino Uno to interact with Lego components, and in doing so, you performed such simple tasks as uploading sketches and wiring up pins. Here, let's delve a little deeper into the technology.

IN THIS CHAPTER

History of the Arduino Project
What is OSHW?
Anatomy of the Uno
The Arduino Ecosystem
Arduino Resources
The Next Chapter

The magic of the Arduino platform is in the size of this ecosystem. Everything from rolling robots to constructs that create art (Figure 3-1) can be—and has been—built with Arduinos.

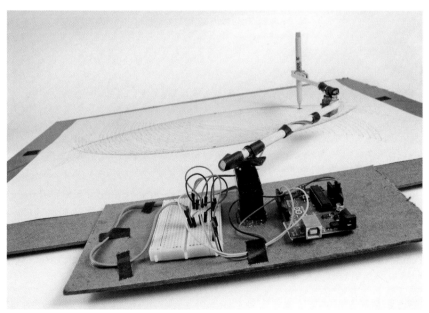

Figure 3-1. *Pete Prodoehl's Arc-O-Matic controls a Sharpie marker with a pair of servos and an Arduino. Credit: Pete Prodoehl*

In this chapter we'll examine the Uno in almost nauseating detail, but we won't neglect Uno's cousin 'Duinos like the Mega and the Fio, and we'll also cover shields, which are Arduino add-on boards. Finally, we'll cover the Arduino language as well as its critical resources of subroutines, global variables, and programming templates called libraries.

History of the Arduino Project

The Arduino project began in 2005 as an educational tool for a design class at the Interaction Design Institute Ivrea, in Ivrea, Italy. The goal was to lure nontechnical people into the realm of electronic design. If an artist seeks to build an interactive sculpture, for example, he or she doesn't necessarily want to learn advanced electrical engineering concepts along the way. Similarly, non-EE students were often shut out of the realm of microcontrollers thanks to the steep learning curve. Therefore, the nascent Arduino team (Figure 3-2) resolved to make their microcontroller inexpensive, easily programmed, and supported by a community of builders.

Figure 3-2. *The Arduino core team, consisting of Gianluca Martino, David Mellis, Tom Igoe, David Cuartielles, and Massimo Banzi. Credit: Massimo Banzi*

By 2007 over 50,000 Arduinos had been sold, and that number had increased to 300,000 within four years. It's the sort of phenomenon that only grows stronger the more people who experience it. All those students and tinkerers publishing their experiments online served to create a massive international community of collaborators creating code and wiring up electronic projects.

By 2012 the Arduino project has continued to evolve and to get more polished. It has its own logo and brand identity, and built its own factory (Figure 3-3) to manufacture boards. It's one of the most exciting and liberating technological advances today.

Figure 3-3. *The Arduino factory in Torino, Italy. Credit: Phillip Torrone*

What Is OSHW?

"Open source hardware (OSHW) is a term for tangible artifacts—machines, devices, or other physical things—whose design has been released to the public in such a way that anyone can make, modify, distribute, and use those things."

> — Opening sentence of the open source hardware definition 1.0 (*http://freedomdefined.org/OSHW*)

The open source hardware community (Figure 3-4) is a group of people who share the designs for creating new physical things. They release the designs, allowing anyone to copy, change, mashup, and learn from their designs. To fall under the endorsed open source hardware definition, the released designs must even have no restrictions to stop other people from making and selling copies of the actual item! Many technical artifacts in our society are protected, both physically and legally, from people finding out how they work. Proponents of open source hardware seek to enable anyone to learn from their designs.

Figure 3-4. *The open source hardware logo; you'll find it on OSHW boards throughout the world*

Sparkfun, a $20M+ electronics company as of 2012, has said that open source hardware allows them to innovate quickly, promotes learning from each other, keeps them on their toes, forces them to stay sharp, and makes for better, lower-cost products (*http://www.sparkfun.com/news/844*).

While there are many similarities between open source hardware and open source software, there are also differences. The marginal production cost (that is, the extra cost to make another copy) of open source software is negligible. Disk storage is cheaper, comparatively speaking, than buying physical things, storing them, and then shipping them out. Also, the enforceability of open source software appears to be relatively clear. Source code is under copyright, and copyrighted materials can be distributed with licenses. The enforceability of open source hardware is less legally clear. This is partially due to the age of the differing communities, but also due to differences in the law. Hardware intellectual property in the US is mostly governed by patents, not copyright, and the laws are not necessarily analogous. However, communities are held together by more than laws.

One member of the open source hardware community, Phillip Torrone, wrote an editorial entitled "The (Unspoken) Rules of Open Source Hardware," in which he lists a few tenets that "the core group of people who have been doing... open source hardware" try to follow in order to "share our work to make the world a better place and to stand on each other's shoulders and not on each other's toes" (*http://blog.makezine.com/2012/02/14/soapbox-the-unspoken-rules-of-open-source-hardware/*).

The list includes crediting each other for their contributions, voluntarily paying royalties to each other, and not "openwashing" or not using the phrase "open source hardware" without releasing the design files.

Chapter 3

Arduino is an excellent example of a successful open source hardware project. The Arduino team has fostered a learning environment, and the Arduino project both saves experienced engineers' time during prototyping and opens the arena of physical computing by making it usable to new people while staying low cost.

Anatomy of the Uno

What exactly are you getting when you buy or build an Arduino? Figure 3-5 shows you the board and breaks out the role every component plays.

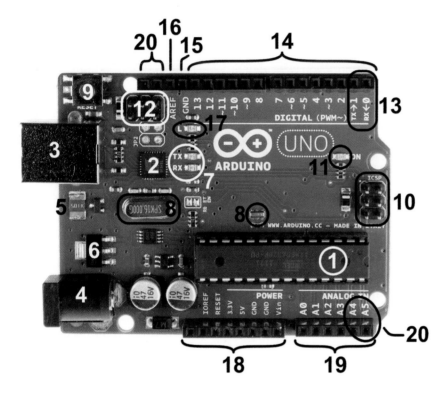

Figure 3-5. *The Arduino Uno, explained. Credit: The Arduino Team*

1. ATmega328P microcontroller: The brains of the Arduino, this is the chip that runs the code you type into the Arduino IDE.

2. ATmega16U2 microcontroller: This little microcontroller runs the USB serial interface between your computer and the Arduino.

3. USB-B jack: Use a USB cable to connect your Arduino to your computer to upload new program code and provide power.

4. 2.1mm center-positive DC power jack: You can also provide power via this DC jack, but the recommended voltage is between 7 and 12 volts.

5. Resettable fuse (PTC): This device limits the amount of current drawn from the USB port to help protect your computer. If there is a problem, it will disconnect the Arduino, and you will need to wait a few minutes for it to reset.

6. 3.3 volt regulator: This little chip takes 5V from the Arduino board and makes 3.3 volts for use by other chips and shields.

7. These LEDs flash every time data is sent or received using the Arduino's serial port pins 0 and 1. When you upload code from your computer to the Arduino, you will see these LEDs flash while the program is being transferred.

8. Crystal oscillators (one for the USB interface chip, the other for the main ATmega that runs your code): This crystal oscillator provides a reliable clock signal for the chips.

9. Reset button: If you want your Arduino code to start over from the beginning, press and release this reset button.

10. ICSP programming header pins: The usual way to program your Arduino is through the USB interface, but if something gets really messed up, you can use a special "hardware programmer" with these pins to fix any problems with your Arduino chip. This is also used for connecting to certain expansion boards (known as *shields*).

11. Power LED: When the Arduino is plugged into the USB port or has a DC power supply plugged in, this LED will light up.

12. ICSP programming header for the ATmega16U2 USB interface chip.

13. Serial UART transmit and receive pins: These pins connect directly to the serial pins on the main chip. They are also connected to the USB interface chip and are used to transfer the program code from the computer to the Arduino.

14. Digital input/output pins: The Arduino code can read and write digital values (0 volts or 5 volts) to and from these pins, and some pins (marked with a ~) can even approximate an analog voltage using Pulse Width Modulation (PWM).

15. Ground pin: There are three ground pin connections on the Arduino Uno, two on the bottom and one on the top.

16. Analog reference voltage pin: This pin can be used to improve the accuracy of the analog-to-digital converter, when measuring an analog (0 volts through 5 volts) signal using the analog input pins.

17. Pin 13 LED: There is an LED connected to pin 13 to help with debugging and testing your Arduino.

18. Power connections: This section of eight pins has an unmarked pin, the voltage reference pin (used by some shields), the reset, 3.3 volts, 5 volts, two ground pins, and the input voltage (Vin) pin, which is connected to the DC jack.

19. Analog input pins: These six pins can measure analog voltages between 0V and 5V, and convert the voltage into a number between 0 and 1023.

20. I2C pins: I2C is a very common way to let your Arduino talk to other small chips, and uses the A4 and A5 pins.

ATMEGA328

The brains of the Arduino Uno is the ATmega328 microcontroller chip, which operates at 20 MHz, features 32KB of flash memory and an 8-bit CPU, 2KB RAM, and 23 I/O pins.

What this really means is that most of the cool stuff from the Uno board is encapsulated in the chip, and when you advance sufficiently in your skills, you can dispense with the Uno altogether and wire the ATmega right into your project.

Case in point, the Defusable Clock (Figure 3-6) from Nootropic Design. Its creator, Michael Krumpus, created a board with the ATmega, an LED display, a couple of buttons,

and some power management features. The idea is that the board is the detonator for a Hollywood-style bomb (it's not a real bomb!) and you must "cut the right wire" to defuse it. The ATmega chooses a random wire to be the correct one, with the other three either do nothing or set off the alarm. Failing to cut the right wire causes a buzzer to make a rude noise while assorted LEDs flash alarmingly.

Mike's project is especially cool because it still functions as an Arduino-compatible platform and can be reprogrammed just like any Arduino, enabling users to customize the project to their own ends.

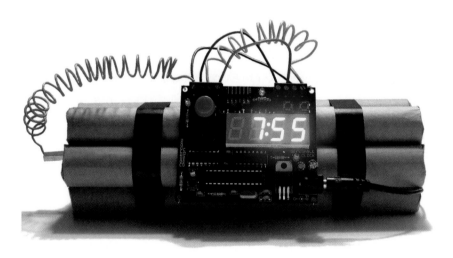

Figure 3-6. *Arduinos pop up in the oddest place; Nootropic Design's Defusable Clock uses an ATmega328 chip running the Arduino firmware. Credit: Nootropic Design*

The Arduino Ecosystem

The Arduino is awesome, no doubt about it. But what really makes the system great is its ecosystem. You don't just buy the board and have to fend for yourself after that. No, you have multiple varieties of Arduino, great add-on boards called shields, and tons of code other people hacked together. By promoting a culture of sharing and support, the Arduino has become the default prototyping and tinkering system for thousands of students, hobbyists, and engineers. The following is just a broad overview of all the options available for you—there is a lot more!

Uno Alternatives

As we've seen from the "ATmega328" sidebar, the heart of the Arduino is its microcontroller chip. Because of that, many different configurations of Arduino can be and have been created. Need a smaller Arduino? Or a bigger one? Chances are it's out there already.

Arduino Mega (Maker Shed P/N: MKSP5, $65)

> The Arduino Mega 2560 (Figure 3-7) is, well, a mega Arduino—it features a more powerful chip (the ATmega2560) with tons more I/O pins. There is another flavor to the Mega, the ADK, essentially the Mega 2560 but with a USB host interface that allows it to connect with Android phones. You'll want to use Megas for those big ol' projects that a single Uno just can't run.

Figure 3-7. *The Arduino Mega gives you more pins and more power! Credit: The Arduino Team*

Arduino Fio (Sparkfun P/N: 10116, $24.95)

> The Fio (Figure 3-8) gives you much the same experience as the Uno, but in a considerably smaller package. In addition to 14 digital I/O pins and 8 analog inputs, it features a built-in resonator. It can be powered with a LiPo battery with charging possible via USB, and has the socket for an XBee wireless module on the bottom of the board, enabling it to be wirelessly reprogrammed.

Chapter 3

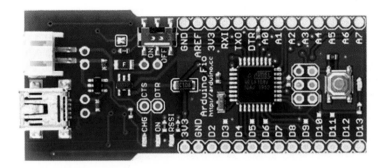

Figure 3-8. *Packing a tiny footprint with many features, the Arduino Fio is a popular choice among hardware hackers*

Arduino Pro Mini (Sparkfun P/N: DEV 11113, $18.95)

The Pro Mini is the culmination of Sparkfun's goal to build the smallest possible Arduino. It lacks many of the bells and whistles that attract hobbyists to the Uno, and is intended for more experienced users (hence the "pro") who don't need the programming headers, voltage regulator, or USB.

Freeduino (Maker Shed P/N: MKSB013, $26)

An Arduino from another mother. Since pretty much everything in the Arduino ecosystem is open source (see "OSHW," earlier in this chapter) pretty much anyone with the skills and motivation can build their own board from the schematics, and even sell it. Electronic kitmaker Solarbotics did just that, and the result is the Freeduino. It's compatible in every way with actual Arduinos but costs a little less.

Diavolino (Evil Mad Scientist Laboratories P/N: 666, $11.95)

A barebones version of the Freeduino, the Diavolino was created by Evil Mad Scientist Laboratories as a low-cost alternative to the Uno. It lacks the USB interface chip and power regulator. On the plus side, it's the perfect size to stick onto a 3xAA battery box, as you can see from Figure 3-9.

Figure 3-9. *Evil Mad Science's Diavolino board, sitting on a battery pack. Credit: Windell H. Oskay*

Lilypad (Maker Shed P/N: MKSF9, $74.95 for the Beginner's Kit)

Based off the Arduino, the Lilypad (Figure 3-10) departs radically from the Uno's paradigm. The Lilypad was designed by Leah Buechley as a *wearable* Arduino. It can be sewn into clothing or accessories, is washable, and may be connected to buttons, LEDs, and sensors with conductive thread. Have an idea for a wearable electronics project? You want the Lilypad.

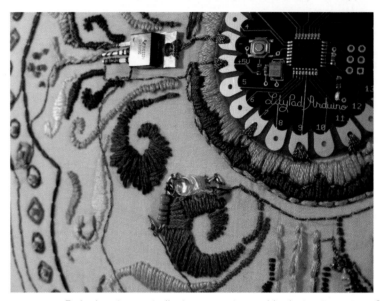

Figure 3-10. *Embed a microcontroller in your next wearable electronics project. Credit: Becky Stern*

Older Models

Not every Arduino floating around represents the latest version. Sometimes all you have to work with is a version previous to the Uno, and guess what? Chances are it'll work fairly well.

Duemilanove

Almost identical to the Uno! The Duemilanove (Figure 3-11) has the same ATmega chip and most everything else matches as well. It has a different USB-to-serial chip and a less pronounceable name, but other than that, you really can't go wrong with a Duemilanove as a substitute for the Uno.

Diecimila

The predecessor to the Duemilanove, the Diecimila was once the go-to board in the misty days of yore—like, 2007. It is largely the same as its successor but with half the RAM and half the memory. However, if your project is not too big, you can totally rock one of these.

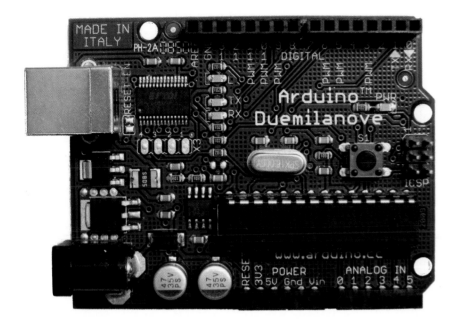

Figure 3-11. *Don't have an Uno? The Duemilanove holds up fairly well*

Shields

Shields are add-on boards for Arduinos. They're usually the same size as an Uno, and feature I/O pins on the underside, allowing you to nest a shield directly on top of the Arduino. If there are plenty of leftover pins, shield builders will often add female pinouts to the shield so another one can be added on top, theoretically allowing you to rock a whole stack of shields. The following are a handful of the more commonplace products available. Want an exhaustive list? Check out *http://shieldlist.org*.

Voice Shield (Maker Shed P/N: MKSKL3, $45.95)

> The Voice Shield adds audio playback to your Arduino projects. Say you want to build a talking robot—you'll want one of these. Plug speakers into the jacks and an Arduino into the underside, and the shield will play back MP3s stored on the chip.

Video Game Shield (Maker Shed P/N: MKWL02, $22.50)

> Wayne & Layne's Video Game Shield (Figure 3-12) allows you to make and play classic black & white video games like Pong and Snake, interfacing with your TV and a pair of Wii Nunchuks.

Figure 3-12. *Wayne and Layne's Video Game Shield plugs into an Arduino, allowing you to play old-school video games*

Motor Shield (Maker Shed P/N: MKSP12, $30)

Motor control is somewhat of a drag with Arduinos, because they use up tons of I/O pins. To the rescue comes the Arduino Motor Shield, which packs an L298 full-bridge motor driver (and cousin to the L293D half-bridges we use in the Bricktronics shield featured in this book!) and can run two stepper or DC motors.

Ethernet Shield (Maker Shed P/N: MKSP7, $54.99)

Connect your Arduino project to the Internet with the help of the Ethernet Shield (Figure 3-13) and its Wiznet W5100 chip. Want to have your project send out tweets or search the web for specific character strings? Get one of these. Like many shields, you can stack another on top of it to make an even more complicated project.

Figure 3-13. *The Ethernet Shield connects your robot to the Internet*

Datalogger Shield (Adafruit P/N: 243, $19.50)

The datalogger shield essentially adds an SD card slot to your project, enabling you to either access data from the card or to write data to it. A real time clock chip (described in Chapter Four) timestamps each entry. Even better, the datalogger's relatively sparse footprint requirements means there's a prototyping area where you can add your own circuitry.

LoL Shield (Maker Shed P/N: MKJR3, $25)

The LoL Shield's name refers to Lots of LEDs, as we can see in Figure 3-14. It's an LED matrix board and each light may be addressed individually, allowing scrolling graphics and messages to be displayed.

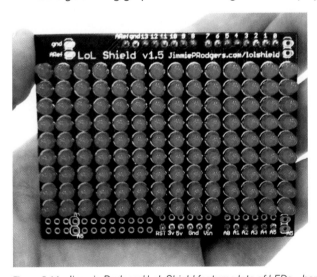

Figure 3-14. *Jimmie Rodgers' LoL Shield features lots of LEDs—hence the name. Credit: Adafruit Industries*

Danger Shield (Sparkfun P/N: DEV-10570, $29.95)

> If the LoL Shield is a board covered with LEDs, the Danger Shield is stuffed full of other stuff you might need: three potentiometers, a 7-segment LED, a temperature sensor, a photocell, a buzzer, and some buttons. The question isn't what you can do with a Danger Shield—ask what you can't do with it!

Arduino Resources

When you buy an Arduino, you're not just buying a board—you're becoming part of a huge gathering of brilliant people, all contributing their knowledge to a vast pool of resources. When you work on a project, chances are someone has already researched your angle, and has published a blog post or written an article about their adventures. The following are some likely books and websites for your enrichment.

Books

The following books are some of our favorite Arduino reference materials:

Arduino Cookbook (by Michael Margolis)

> This massive reference (Figure 3-15) is many Arduino hackers' favorite thanks to the huge number of obscure and difficult projects the book tackles. It's definitely geared more toward the experienced tinkerer rather than the newbie, but even a newcomer will eventually get a lot of information out of the book.

Getting Started with Arduino (by Massimo Banzi)

> The perfect pocket guide (Figure 3-16) for new Arduino fans, it was written by one of the project's co-creators. It's a wee book but loaded with info, projects, and tutorials for those dipping their toe into the Arduino pool.

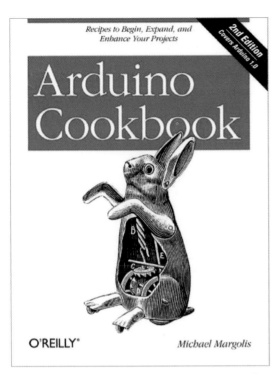

Figure 3-15. *The Arduino Cookbook features wiring diagrams and code examples for dozens of commonplace and obscure Arduino projects*

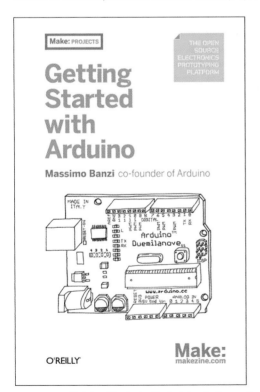

Figure 3-16. *Getting Started with Arduino is the perfect beginner's reference to the Arduino phenomenon*

MAKE Magazine

The ultimate DIYer quarterly, MAKE (Figure 3-17) delves into dozens of awesome projects with every issue. There are always a few Arduino projects in every issue, but the other articles are very informative as well.

Figure 3-17. *MAKE Magazine is the de facto journal of the maker movement*

Make: Arduino Bots and Gadgets (by Kimmo Karvinen and Tero Karvinen)

Build six fun Arduino projects including an insect bot and an interactive wall painting. The authors also show how to control your creations with your smart phone! Figure 3-18 shows the cover.

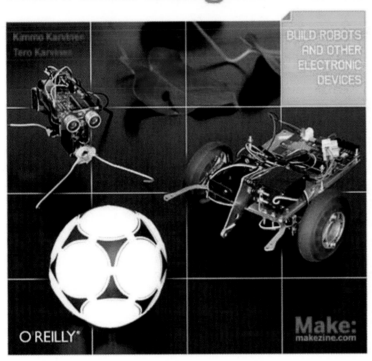

Figure 3-18. *Make: Arduino Bots and Gadgets features six Arduino projects you can build*

Making Things Move
by Dustyn Roberts

> Basic mechanical principles for non-engineers, illustrated with an assortment of sample projects.

Making Things See
by Greg Borenstein

> All about hacking the Kinect, Microsoft's cool new game controller that can actually see you move.

Making Things Talk
by Tom Igoe

> Arduino core team member Igoe's encyclopedic take on creating electronic projects that interact with the physical world.

Websites

While reference books are important—especially the one you're reading right now!—Internet sites like blogs have the advantage of being frequently updated. Need the latest Arduino info? Check out one of these sites.

Adafruit Industries
http://adafruit.com

> Ladyada and confederates run Adafruit Industries, a DIYer's dream site with tons of tutorials as well as kits and electronic components for sale. Also be sure to check out Ladyada's Arduino tutorial: *http://www.ladyada.net/ learn/arduino/*.

Arduino
http://arduino.cc

> The motherlode of Arduino information: the project's home page. Also be sure to check out the Arduino Playground—*http://arduino.cc/playground/*—the ultimate Arduino wiki.

Bildr
http://bildr.org

> DIYer blog focusing on how-to articles instead of projects.

Evil Mad Scientist Laboratories
http://evilmadscience.com

> EMSL are Windell Oskay and Lenore Edman, makers dishing out info and innovative open-source creative and electronic projects.

Freeduino
http://Freeduino.com

> The home site of the Freeduino project has tons of resources for noobs and experts.

Instructables
http://instructables.com

> This site features clear step-by-step instructions for countless types of endeavors. Find their Arduino projects here: *http://www.instructables.com/technology/arduino/.*

MAKE
http://makezine.com

> The ultimate DIYer resource, with thousands of articles exploring every facet of the maker community. Their Arduino page may be found at *http://makezine.com/arduino.*

Make: Projects
http://makeprojects.com

> MAKE Magazine's project wiki, featuring step-by-steps on everything from pyrotechnics to moss graffiti. A lot of projects from older MAKE and CRAFT magazines ended up here, score!

Maker Shed
http://makershed.com

> The companion store to the MAKE ecosystem, with kits, books, and apparel celebrating the maker ethos.

Code

The easiest way to learn to program is to study the architecture of existing sketches. Need to learn how to do something? Chances are it has been done already. In a way, this is the ultimate expression of the value of the Open Source movement—you gotta buy your hardware, but the code is free!

Arduino Code Examples
http://arduino.cc/en/Tutorial/HomePage

> This site serves up the ultimate collection of code—both sketches and libraries—for Arduinos.

Arduino Playground
http://www.arduino.cc/playground/Main/Resources

> The Playground is the publicly editable wiki for the Arduino project. It's all here! Just look for it.

The Next Chapter

In Chapter Four we return to our robot making with our Clock model. It's an analog Mindstorms clock that displays the correct time and offers intriguing expansion possibilities like a settable alarm. It's pretty cool and we know you'll enjoy building it.

Project: Clock

4.

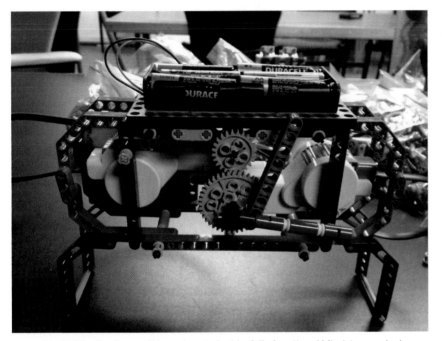

Figure 4-1. *In this chapter you'll learn how to build a fully functional Mindstorms clock*

Our next project is a simple one—or so we imagined when we began designing it. We thought, "What could be simpler than a Lego clock? Sounds like a great project for a book about Lego and Arduino projects." Stick a couple of motors on a Lego frame and have the Arduino move the motors to tell us what time it is.

Fortunately, it wasn't that simple, and we ended up adding a lot of complexity to the project. We built this project before we created the Bricktronics Shield, so we had to figure out how to control motors and display the time via clock hands (see "About the L293D Chip"). We say "fortunately" because we learned an insane amount about interweaving the two systems, and we're sharing what we learned with you. Read on!

The clock hands move with the help of two standard Mindstorms motors, controlled by an Arduino Uno, and a couple of buttons allow you to set the time.

A NOTE ABOUT MODDING LEGO

Lego fanatics will howl when they see what we had to do to the 40-tooth gear in order to make this design work. To AFOLs (Adult Fans of Lego), damaging, modifying, or gluing Lego elements is considered anathema, as is using non-Lego parts in your model. Sorry, folks! This is just the second project and we're already altering and gluing Lego and using Arduino in place

of the familiar NXT microcontroller bricks. Our philosophy is that it's okay to mod your bricks in any way that meets your needs, just like any other tool. By adding Arduino, our readers will learn a lot about how Mindstorms *and* electronics work. We call that a win, even if it troubles purists along the way.

Parts List

The following tools, electronic components, and Lego parts will be needed to complete the clock:

Tools & Electronics

- Arduino Uno
- Bricktronics Shield
- 4 Mindstorms wires
- Power supply rated for 9V at 1.3A or greater with a 2.1mm center-positive plug. This provides power to the Arduino and Lego motors.
- Drill & 3/16th inch drill bit
- Hot glue gun

ABOUT THE L293D CHIP

For the original iteration of these projects, we chose the L293D motor-control chip, and we went on to use that as the heart of the Bricktronics Shield. You'll be using the Bricktronics shield in this project, but here's some background on the L293D.

The L293D is a robust motor control chip that enables precise control of one or two motors including voltage and direction. Let's examine both of these factors. First, the controller can tell a servo whether to turn forward or backward with the help of three inputs: A, B, and Enable. If Enable isn't active—in electronics parlance, we call this *high*—the motor doesn't turn. When Enable and A are high, but B is low, the motor turns one way at full speed. When Enable and B are high, while A is low, the motor turns the other way at full speed. If both A and B are at the same level, the motor doesn't turn.

Second, we can also use these inputs to control the speed of the motors, through *pulse width modulation* (PWM)—

switching between two voltages fast, creating a series of pulses. The length of the high part of the pulse compared to the length of the low part of the pulse determines the effective "value" of the PWM voltage. In Arduino, the value can range between 0 and 255. 255 means maxed out at 5V, while 0 means no voltage. If we put a PWM voltage into B, while keeping A low, we can change the speed of the motor. When the PWM pulse is high, the motor will move, and when it's low, the motor won't be moving, but these happen so fast that it is mostly unnoticeable.

So, why do we need the L293D? Can't the Arduino drive motors? The quick answer is yes, an Arduino can control hobby servo motors quite well, but the Mindstorms motors are more complicated and need a dedicated motor control chip to avoid maxing out or damaging Arduino pins. For a longer and more detailed answer, see Chapters Two and Three where we delved into Mindstorms and Arduino, respectively.

Figure 4-2. *The L293D is a handy chip if you want to drive motors with an Arduino*

Lego Elements

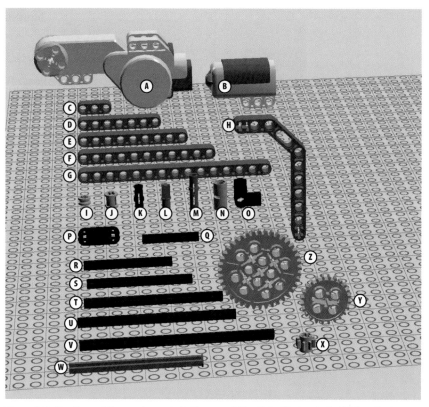

Figure 4-3. *You'll need these elements in the quantities listed below to build the clock*

A. 2 Mindstorms motors

B. 2 touch sensors

C. 4 3M Technic beams

D. 2 7M Technic beams

E. 1 9M Technic beam (blue or some other bright color*)

F. 6 11M Technic beams

G. 4 15M Technic beams*

H. 12 double angle beams, 3x7*

I. 11 half bushes (2 of them yellow)

J. 23 bushes

K. 26 connector pegs

L. 28 cross connectors

M. 6 3M connector pegs

N. 7 tubes (3 of them brightly colored)*

O. 1 90-degree angle element

P. 2 double cross blocks

Q. 1 4M cross axle

R. 2 6M cross axles

S. 2 7M cross axles

T. 2 9M cross axles

U. 5 10M cross axles*

V. 1 12M cross axles

W. 1 8M cross axle with end stop

X. 2 8-tooth gears*

Y. 3 24-tooth gears*

Z. 1 40-tooth gear* (modified; see "Prepare the Gear," later in this chapter)

* Not found in the Mindstorms set at all or in insufficient quantities. See "A Note About Sourcing Lego" to learn what you can do.

Assembly Instructions

Once you have gathered together the parts, it's time to build! The following instructions show you how to build the clock. While you put it together, feel free to experiment and alter the design—the worst that could happen is that the clock just won't work! If that happens, just back up to the previous step and start over again.

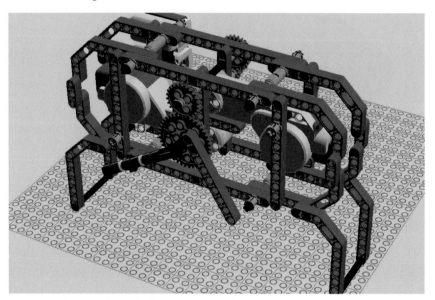

Figure 4-4. *The completed clock model—just add electronics*

Prepare the Gear

The only tricky part of the Lego build is that you'll have to modify the central 40-tooth gear in order for the clock to work. The reason is that in order for the hour and minute hands to both rotate freely without fear of collision, the gear upon which the hour hand is attached must rotate freely around the axle that turns the minute hand. Unfortunately, Lego doesn't make a gear with a smooth central hub. Their gears all feature cross-axle hubs, which makes sense—you usually want your gear to derive its energy from a cross axle, or impart energy to one.

Since Lego has not seen fit to create such a gear, we did. We took a 40-tooth gear and filled in the holes around the axis with hot glue, in order to strengthen the hub for drilling. Then, using a 3/16th-inch bit and power drill, we drilled out the hub so it rotates freely around a standard cross axle.

Figure 4-5. *We modified the 40-tooth Technic gear to have a smooth hub; We filled in the surrounding holes with hot glue to ensure the drilling doesn't weaken the gear*

Build the Lego Model

Next, we'll assemble the clock itself. This is your opportunity to experiment with the model's design and customize it to your own liking.

1. Begin with a couple of 3x7 double angle beams as shown in Figure 4-6.

2. Add two 15M beams and one 11M beam (see Figure 4-7).

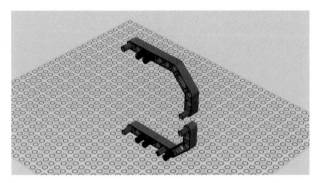

Figure 4-6. *Step 1: Setting up the angle beams*

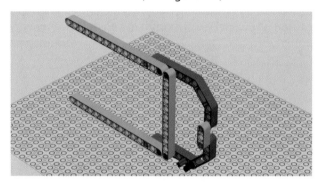

Figure 4-7. *Step 2: Adding some more beams*

3. Now work on the other side with two more 3x7 double angle beams, as shown in Figure 4-8.

4. You'll need supports as well; add one 3M and one 11M beam. Figure 4-9 shows how they go together.

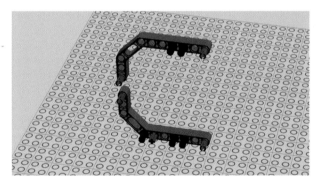

Figure 4-8. *Step 3: Making the other side with angle beams*

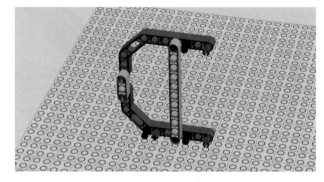

Figure 4-9. *Step 4: Adding two supports*

Chapter 4

5. Combine the two assemblies (Figure 4-10) and set them aside for now.

6. Next, connect two pins to an 11M beam as shown in Figure 4-11.

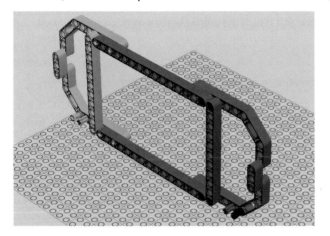

Figure 4-10. *Step 5: Combining the two assemblies*

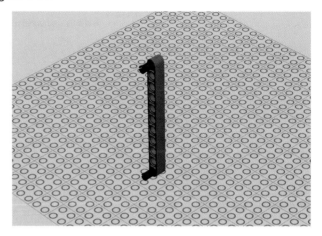

Figure 4-11. *Step 6: Connecting the pins to the beam*

7. Next, connect an 11M beam (see Figure 4-12).

8. Take what you have so far (Figure 4-13) and build another one just like it!

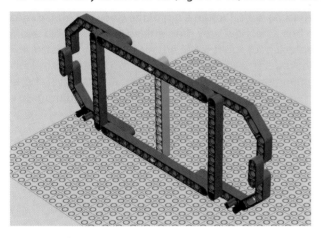

Figure 4-12. *Step 7: Adding the beam from the previous step*

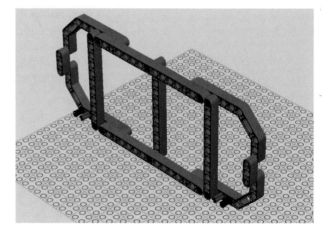

Figure 4-13. *Step 8: Make one just like this*

9. Add four 10M cross axles and bushes to one of the assemblies you built as shown in Figure 4-14, but leave the other one alone for now.

10. Add half bushes to the back to keep the cross axles in place as shown in Figure 4-15. (By the way, don't use the yellow half-bushes we specified; you'll need them for the clock's hands.)

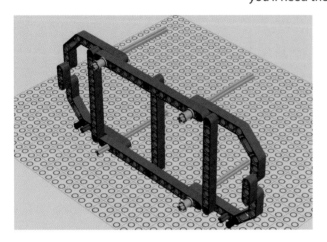

Figure 4-14. *Step 9: Adding 10M cross axles and bushes*

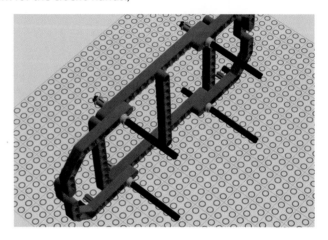

Figure 4-15. *Step 10: Adding half bushes*

11. Add four pipes and four more half bushes. Figure 4-16 shows how it goes together.

12. Add some 9M cross axles and bushes (Figure 4-17).

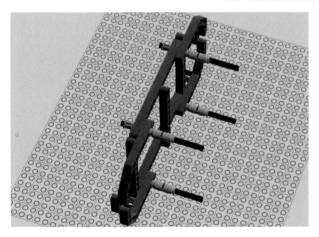

Figure 4-16. *Step 11: Adding pipes and half bushes*

Figure 4-17. *Step 12: Adding more cross axles and bushes*

Chapter 4

13. Now let's work on the motors. Add three 3M connector pegs as shown in Figure 4-18.

14. Connect a 7M Technic beam (see Figure 4-19).

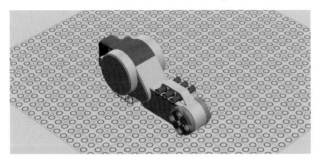

Figure 4-18. *Step 13: Add connector pegs to the motors*

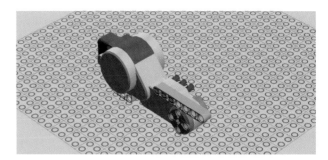

Figure 4-19. *Step 14: Connecting the beam*

15. Grab the second motor, and add three 3M pegs as shown in Figure 4-20.

16. Connect the two motors together with the help of the Technic beam (see Figure 4-21).

Figure 4-20. *Step 15: Adding pegs to the other motor*

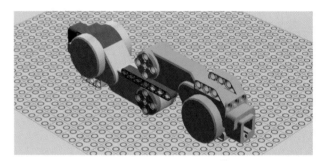

Figure 4-21. *Step 16: Connecting the two motors*

17. Throw another 7M beam on there (Figure 4-22).

18. Add the motors to the structure you built earlier, as shown in Figure 4-23.

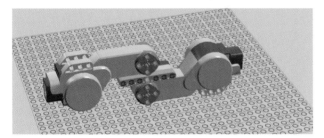

Figure 4-22. *Step 17: Adding another beam*

Figure 4-23. *Step 18: Combining the motors and the assembly*

19. Throw the duplicate assembly onto the back (see Figure 4-24).

20. Add six bushes as shown in Figure 4-25.

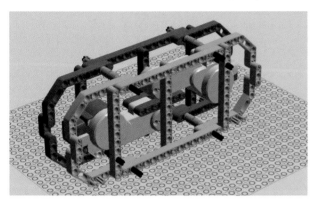

Figure 4-24. *Step 19: Attaching the duplicate assembly*

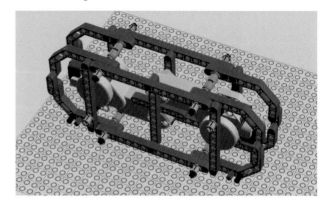

Figure 4-25. *Step 20: Adding bushes*

21. Add angle beams for legs. Figure 4-26 shows this.

22. Insert two 7M cross axles as shown in Figure 4-27.

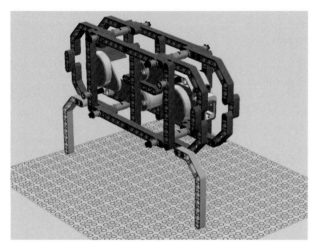

Figure 4-26. *Step 21: Adding the legs*

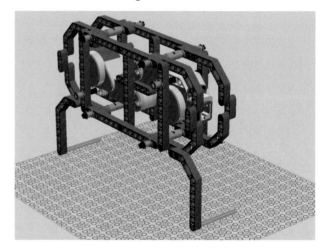

Figure 4-27. *Step 22: Inserting the cross axles*

23. Two more angle beams complete the legs (Figure 4-28).

24. Let's work on gears. Add a 24-tooth gear to the end of a 12M cross axle (see Figure 4-29).

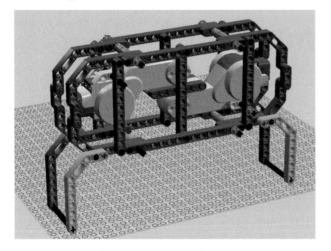

Figure 4-28. *Step 23: Adding more angle beams*

Figure 4-29. *Step 24: Connecting the 24-tooth gear to a cross axle*

25. Thread the cross axle through the center hole, between the motors' orange hubs. Figure 4-30 shows where to put it.

26. Build another gear and axle assembly, this one with a couple of bushes, as shown in Figure 4-31.

Figure 4-30. *Step 25: Threading the cross axle*

Figure 4-31. *Step 26: Building another gear and axle assembly*

-

27. Insert the cross axle through the topmost visible hole in the vertical support (Figure 4-32).

28. Secure the axles with two bushes each as shown in Figure 4-33.

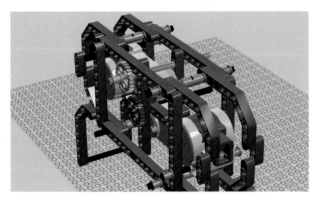

Figure 4-32. *Step 27: Inserting the cross axle through the top*

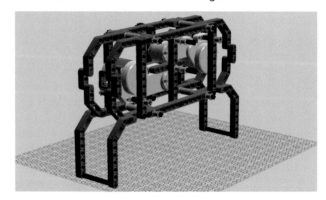

Figure 4-33. *Step 28: Securing the axles*

29. Add two bushes and an 8-tooth gear to a 6M cross axle. Figure 4-34 shows how they go together.

30. Connect the assembly from Step 29 to the upper motor's hub so that the top two gears mesh (Figure 4-35).

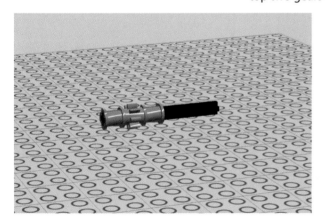

Figure 4-34. *Step 29: Adding the bushes and 8-tooth gear to a cross axle*

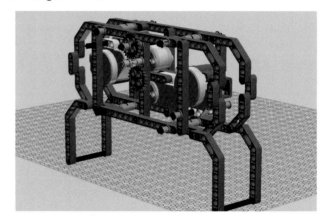

Figure 4-35. *Step 30: Connect to the upper motor's hub*

31. Add a bush and an 8-tooth gear to a 4M cross axle as shown in Figure 4-36.

32. Connect the cross axle to the lower motor's hub, and make sure the lower two gears mesh. Figure 4-37 shows how it all connects.

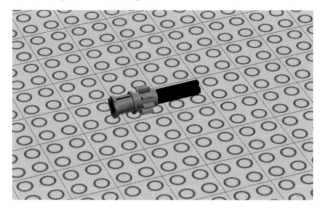

Figure 4-36. *Step 31: Adding a bush and 8-tooth gear to a cross axle*

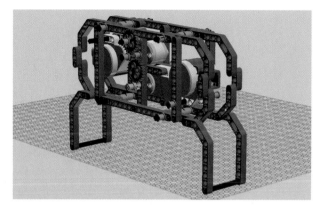

Figure 4-37. *Step 32: Connecting the cross axle to the lower motor's hub*

33. Add a 24-tooth gear and half-bush (Figure 4-38).

34. Add the drilled out 40-tooth gear. Make sure it rotates freely, then mesh it with the top gear as shown in Figure 4-39.

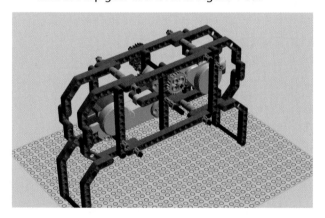

Figure 4-38. *Step 33: Adding a 24-tooth gear and half-bush*

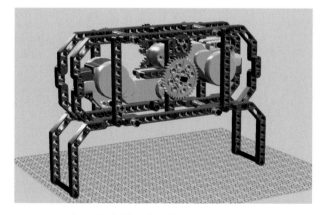

Figure 4-39. *Step 34: Adding the 40-tooth gear*

35. Add the angle element and two pegs (Figure 4-40).

36. Connect a 9M beam to the 40-tooth gear (see Figure 4-41). This is your hour hand.

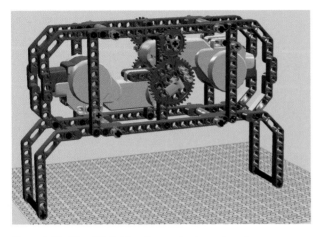

Figure 4-40. *Step 35: Adding the angle element and pegs*

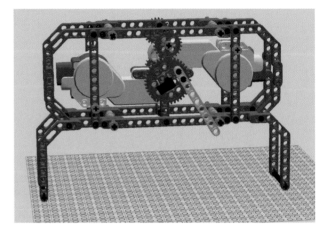

Figure 4-41. *Step 36: Building the hand*

37. Build the minute hand out of a cross axle with end stop, three pipes, and two half bushes. Figure 4-42 shows the assembly.

38. Add the minute hand to the angle element as shown in Figure 4-43.

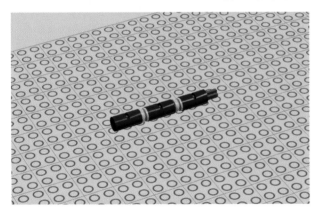

Figure 4-42. *Step 37: Building the minute hand*

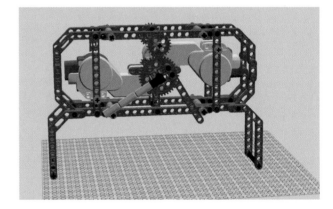

Figure 4-43. *Step 38: Attaching the minute hand*

39. Next, let's work on the button assembly. Add two double cross blocks to a pair of 6M cross axles (Figure 4-44).

40. Add two touch sensors to the cross axles as seen in Figure 4-45.

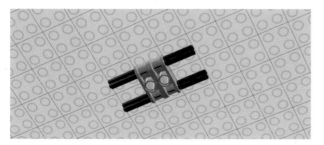

Figure 4-44. *Step 39: Adding cross blocks*

Figure 4-45. *Step 40: Attaching touch sensors*

41. Secure the ends of the cross axles with bushes (see Figure 4-46).

42. Add a couple of connector pegs (Figure 4-47).

Figure 4-46. *Step 41: Adding bushes*

Figure 4-47. *Step 42: Attaching two pegs*

43. Add a couple of connector pegs to attach the buttons (Figure 4-48).

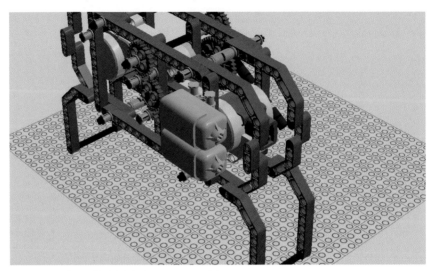

Figure 4-48. *Step 43: Add the button assembly*

Install the Arduino

As with the Drawbot in Chapter One, you'll need to use Bricktronics mounting plates to attach your Arduino to the clock, as shown in Figure 4-49. Use the exposed cross connectors and cross axle ends on the back of the clock and thread them through the Technic holes in the plates, just like you did in Chapter One. Attach wires per Figure 4-50. Power the clock with an Arduino-compatible wall wart (9-12V, 2.1mm center positive barrel connector, 1.3A or more) to run the clock. All you have to do is program the clock and you're done!

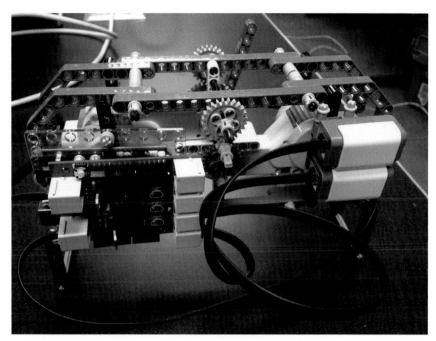

Figure 4-49. *Use your Bricktronics mounting plates to add the Arduino*

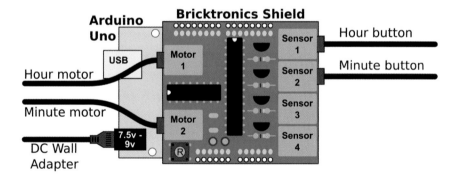

Figure 4-50. *Attach the Mindstorms wires as you see here*

Program the Robot

Once you've built the clock, it's time to program it! Plug in your Arduino and upload the program code found in the Bricktronics library (found on *http://wayneandlayne.com/bricktronics* if you haven't snagged it already!)

The clock code follows a general form which is seen throughout this book. First, information about the environment is gathered. Second, we determine a reaction. Third, we react until we gather more information about the environment.

We do this in the clock with a main loop that only iterates every hundred milliseconds or so, and if the code in the loop executes faster than that, it idles until the next iteration. We've found this to be a useful code framework for mechatronics.

The libraries used are the Time library, which brings POSIX-style time structs to Arduino, and the Bricktronics libraries.

The main loop checks to see if the time-setting buttons have been pushed, and then it calculates where the minute and hour hands should be. It sets those positions using the PID calls, and waits for the next main loop.

```
#include <Wire.h>
#include <Adafruit_MCP23017.h>
#include <Bricktronics.h>
#include <Time.h>

#define TIME_STEP 100
time_t t;

Bricktronics brick = Bricktronics();
PIDMotor h = PIDMotor(&brick, 1);
PIDMotor m = PIDMotor(&brick, 2);
Button hour_button = Button(&brick, 1);
Button minute_button = Button(&brick, 2);

void setup()
{
  Serial.begin(115200); ❶
  Serial.println("starting!");

  brick.begin();
  m.begin();
  h.begin();
  hour_button.begin();
  minute_button.begin();
}

void digitalClockDisplay() ❷
{
  Serial.print(hour());
  printDigits(minute());
  printDigits(second());
}
```

Chapter 4

```
void printDigits(int digits) ❸
{
  Serial.print(":");
  if (digits < 10)
  {
    Serial.print('0');
  }
  Serial.print(digits);
}

void increment_minute()
{
  adjustTime(60);
  Serial.println("Advance minute pressed");
}

void increment_hour()
{
  adjustTime(3600);
  Serial.println("Advance hour pressed");
}

void check_buttons() ❹
{
  static char last_minute_status = HIGH;
  static char last_hour_status = HIGH;

  char minute_status = minute_button.is_pressed();
  char hour_status = hour_button.is_pressed();

  if (minute_status == HIGH && last_minute_status == LOW)
  {
   increment_minute();
  }

  if (hour_status == HIGH && last_hour_status == LOW)
  {
   increment_hour();
  }

  last_minute_status = minute_status;
  last_hour_status = hour_status;
}


void loop() ❺
{
  long next_loop = millis() + TIME_STEP; ❻
  check_buttons(); ❼
  t = now(); ❽
```

```
int full_hours = hour(t) % 12;
int minute_position;  ❾
minute_position = minute(t)*6 + 360*full_hours;  ❿ ⓫
int m_pos;
m_pos = -3 * 2 * minute_position;  ⓬

int hour_position;  ⓭
hour_position = (hour(t) % 12) * 30 + minute(t)/2;  ⓮
int h_pos = 2 * 5 * hour_position;  ⓯
digitalClockDisplay();
Serial.println();

m.go_to_pos(m_pos);
h.go_to_pos(h_pos);⓰

while (millis() < next_loop)
{
  h.update();
  m.update();
  delay(50);
}
}
```

❶ We use the USB serial port for debugging our Arduino code. It allows us to send information from the Arduino to the computer.

❷ `digitalClockDisplay()` prints out the time to the serial port.

❸ `printDigits()` is a helper function for `digitalClockDisplay()`.

❹ `check_buttons()` handles most of the logic of the clock, outside of moving the hands. It is only called once each time `loop()` runs, which is every 100 ms, so *debouncing* (adding additional program logic to handle electrical noise in button presses) the buttons isn't really necessary in this application.

❺ `loop()` runs over and over again. It handles moving the hands, as well as reading the buttons.

❻ `next_loop` is used for the inner PID loop, which runs for `TIME_STEP`.

❼ `check_buttons()` checks the minute and hour buttons for presses.

❽ t holds the current time in a `time_t` struct.

❾ `minute_position` is the desired position of the minute hand (in degrees) for the current time, where 0 degrees is at the 12:00 position and increases in a clockwise manner.

❿ To convert from minutes to degrees on a clock, you multiply by 6. For example, 15 minutes*6 = 90 degrees.

⓫ `m_pos` is the position we want the minute hand motor to go to. Because of gearing, this is not just `minute_position`!

Chapter 4

⑫ The negative sign is here because of the direction the motor is mounted in. The 3 is based on the gear ratio: there is an 8-tooth gear on the motor, geared to a 24-tooth gear which is connected to the hand. This is a 3:1 ratio, so we need to move our motor three degrees for every one degree we want the minute hand to go. The 2 is because we read two steps per degree.

⑬ `hour_position` is the desired position of the hour hand (in degrees) for the current time, where 0 degrees is at the 12:00 position and increases in a clockwise manner.

⑭ To get the position of the hour hand, we have to move the hand 30 degrees for every full hour, and then move the hour hand proportionally for how far the minute hand has gone. So, for example, 6:00 would work out to be exactly 180 degrees (hour hand pointing at the 6). And at 1:30, the hour hand will be at exactly 45 degrees.

⑮ There is an 8-tooth gear driving a 24-tooth gear, for a 3:1 ratio and then that 24-tooth gear is driving a 40-tooth gear, for a 1.667:1 ratio. This works out to a 5:1 ratio overall. The reason for the 2 in the equation is that we read two steps per degree.

⑯ The program will continue to try to get the hands to the right positions for about 100 ms.

Setting the Clock

The clock keeps accurate time, but it doesn't start off *knowing* the time. The clock's code assumes that it's beginning at 12 o'clock, so you'll need to set it accordingly. First, plug in the clock and hit the minute and hour buttons to move the hands to 12:00, if they're not there already. Then unplug the clock and plug it back in, and you'll be ready to use the buttons to set the clock to its true time.

The Next Chapter

In Chapter Five we'll tackle what is perhaps the most complicated robot in the book, a machine that actually prepares a glass of chocolate milk at the touch of a button! What's not to like about that?

Project: Chocolate Milk Maker

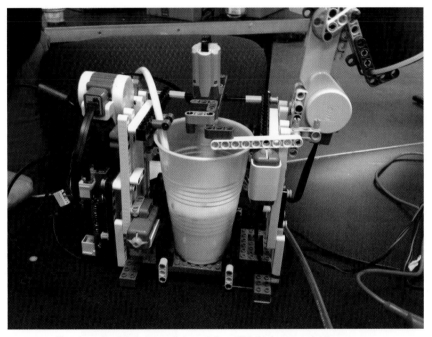

Figure 5-1. *Need a refreshing glass of chocolate milk? Just press a button*

For our next project we're going to build a robot that makes a cup of chocolate milk by mixing together chocolate syrup and regular milk. It's silly, of course— who needs such a thing? Still, it's a great opportunity to expand our knowledge of Lego robotics and to introduce a couple of new tools. Plus, everyone loves a treat!

The way the robot works is that it pumps milk from its jug to a beverage cup using an aquarium pump, which pressurizes the jug and makes the milk to squirt out through an outlet tube, which conveniently dispenses into a glass. While the pump moves the milk, a second assembly squeezes a 6-oz. bottle of chocolate syrup while a stirring assembly mixes up the drink. All of this happens at the touch of the button.

This is the most complicated project in the book from a mechanical stand-point. Everything from the chocolate-squirting assembly to the stirring arm went through numerous iterations, and one thing we discovered was that it was easy to make "bad" chocolate milk, but making delicious chocolate milk means a lot of tweaking. Much of it is subjective. How dark do you like your chocolate? How important is it that the mix is stirred up thoroughly? These are the questions only you can answer. Either way, you're likely to want to modify the code to make it work the way you like.

Parts List

You'll need a wide, and rather odd, selection of parts to build the Chocolate Milk Maker. Never fear, we've listed everything you'll need. As is the case elsewhere in the book, feel free to substitute if you don't have a specific part.

Tools & Electronics

Figure 5-2. *You don't have to pump liquid to move it—just displace it with air; we used this battery powered aquarium pump, shown with the battery cover removed*

Chapter 5

- Arduino Uno
- Bricktronics Shield
- Mounting plate for Arduino
- Tape or hot glue gun
- Battery-powered aquarium pump (we used Marina P/N 11134) (Figure 5-2)
- Zip ties (we used a pack of assorted zip ties, Gardner-Bender P/N 50398)
- Drill & 1/4" bit
- Screwdriver & some #24 0.5" wood screws
- Molex pins, P/N 22-01-3047
- Molex 4 pin terminal housing, P/N 22-01-3047

Beverage Handling

- Wilton 1904-1166 6-oz. food-safe squirt bottle
- Clear Tygon B-44-3 Beverage Tubing (Figure 5-3) with a 3/16" inner diameter, 1/4" outer diameter, and a 1/32" wall
- 9-oz. Hefty cups (Hefty P/N C20950). We designed our robot chassis to accommodate this cup's dimensions. If you go with a different cup, you'll need to rework the robot.
- A rigid container for holding the milk. One bottle we tried was a 1-liter water bottle, a H20 Junior (Arrow Plastic, P/N 65134).
- A plastic spoon; any standard-size spoon will work.

Figure 5-3. *Make sure to use food-safe tubing! We bought this Tygon tubing on Amazon*

Food

Chocolate syrup

> We used Nestlé Quik, which is less viscous than Hershey's. You probably won't be able to get this design to work with the variety of Hershey's that comes in a tin can because it's very thick.

Milk

> We found no difference between the various types of milk. Use whatever kind you like!

A Note About Food Safety

Warning

Any time you make a machine that dispenses beverages you'll want to be absolutely certain no one gets sick. That's why we mandate food-safe plastic for every part that touches milk or chocolate—chiefly the tubing and squeeze bottle. The most notable NON food-safe element in the project is the pump. What we do to skirt the problem is to pump air only, which displaces the milk in the container and forces it out and into the cup. That way we're effectively pumping the milk without exposing the pump to food!

Lego Elements

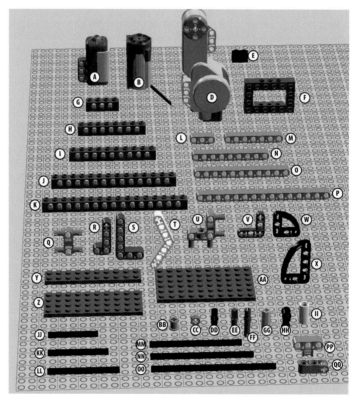

Figure 5-4. *You'll need the following parts to make this robot*

A. 2 Mindstorms touch sensors

B. 1 Power Functions motor

C. 1 Power Functions extension wire (P/N 8886, modified as per Figure 5-97, not pictured in Figure 5-4)

D. 2 Mindstorms motors

E. 2 Technic brick 1x2 with cross hole

F. 2 Technic brick 4x6

G. 4 Technic brick 4M

H. 14 Technic brick 8M

I. 3 Technic brick 10M

J. 8 Technic brick 14M

K. 2 Technic brick 16M

L. 14 Technic beam 3M

M. 11 Technic beam 7M

N. 4 Technic beam 9M

O. 1 Technic beam 11M

P. 11 Technic beam 15M

Q. 2 Technic beam 3M w/4 snaps

R. 1 Technic angle beam 4x2

S. 5 Technic angle beam 3x5

T. 1 Technic angle beam 4x4

U. 4 90 degree angle beam w/4 snaps

V. 4 3x3 Technic lever arms

W. 2 halfbeam curves 3x3

X. 2 halfbeam curves 3x5

Y. 3 plates 2x10

Z. 2 plates 4x10

AA. 1 plate 6x10

BB. 10 cross-axle bushes

CC. 12 half bushes

DD. 79 connector pegs

EE. 25 connector pegs w/cross axle

FF. 13 3M connector pegs

GG. 2 cross axle extensions

HH. 2 catches w/cross hole

II. 2 tubes

JJ. 1 cross axle 5M

KK. 1 cross axle 6M

LL. 1 cross axle 7M

MM. 1 cross axle 9M

NN. 1 cross axle 10M

OO. 1 cross axle 12M

PP. 2 cross blocks 3M

QQ. 1 cross block 3x2

Assembly Instructions

As you build your Chocolate Milk Maker, don't hesitate to play around with the configuration. Maybe you don't like our squeezing mechanism, or you want to try your hand at a different sort of pump, or you want to dispense into a larger glass. Do it! On the other hand, if you just want to build a drinkbot, just follow these steps to get a robot that dispenses a refreshing glass of chocolate milk with the press of a button!

Build the Pump Assembly

First, we'll build the assembly that draws milk out of the jug and pours it into the cup. Though we're tempted to call the entire assembly the pump, more properly only the aquarium pump deserves that term. The milk jug is called an "air-pressurized reservoir." It's similar in many ways to a Super Soaker squirt gun, which you pump to pressurize the water tank to make the water squirt out. In our case, we don't need to squirt milk thirty feet—just far enough to get it into the cup.

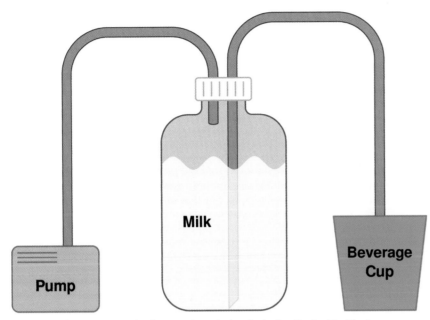

Figure 5-5. *The air-pressurized reservoir works by displacing the liquid with air*

First, prepare the reservoir (a.k.a., the milk jug) by drilling two holes in the top (Figure 5-6) with diameters conforming to the tubing you chose. Next, grab your tubing. We used clear Tygon beverage tubing with a 3/16" inner diameter, 1/4" outer diameter, and a 1/32" wall. You can buy this from Amazon for about a dollar a foot. It's food-safe and great! Another cool thing about this product is that it's available in a bunch of different diameters and wall thicknesses, so you can tailor the tubing to the unique needs of your project.

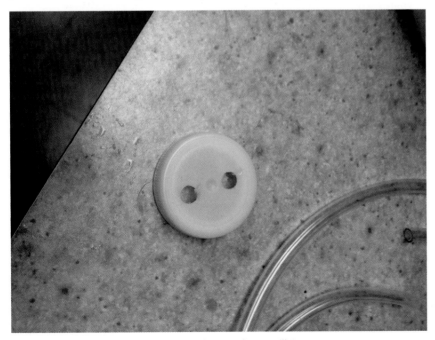

Figure 5-6. *Drill two quarter-inch holes in the top of your milk jug*

Project: Chocolate Milk Maker

Cut off one length of tubing and give it a diagonal cut at the bottom so it doesn't get blocked. This is the output tube. Slide it all the way to the bottom of the milk jug. The other tube should be shorter, and shouldn't reach the milk level. This is the input tube, which pumps air into the jug and thereby forces milk into the output tube. The tubes should be snug, and ideally should be hermetically sealed with tape or hot glue to prevent pressure loss.

Two final notes about the reservoir: first, you should choose your jug carefully. We found standard plastic milk jugs to be too soft—their walls expanded, almost balloon-like, as the pump pressurized them. This meant that even after the pump stopped, the reservoir remained pressurized and kept squirting milk. We found that a more rigid container did a better job of not taking on too much pressure. If you can't find a milk jug that works for you, try the BPA-free water bottle we spec in the parts list.

The second thing to keep in mind is that your reservoir assembly is very happy to siphon, so be sure to keep it lower than the robot to minimize this factor.

Wiring the Aquarium Pump

When we chose our aquarium pump, we went for a battery-powered model because we intended to control it electrically (Figure 5-7) with the help of our Arduino plus a transistor to manage it.

First, separate the battery compartment's negative terminal from the ground of the pump. Then attach two wires to the pump, one to the battery compartment's negative terminal, and the other to the ground of the pump. We used a TIP120 Darlington transistor to connect these electrically. The transistor is able to connect and disconnect the batteries to the pump, starting or stopping it. The Arduino controls the transistor. The details of the connections are shown in Figure 5-94.

This option isn't advised if you choose a pump that runs off wall current. If you do that, we recommend either using something like a PowerSwitch Tail (featured in Chapter Nine), or creating a Lego assembly, powered by an NXT motor, that presses the pump's power switch.

Either way, be sure you have the power switch in the "on" position! If the switch is in the off position, we can't control it with the transistor. (We spent about 15 minutes "debugging" that one!) Good luck!

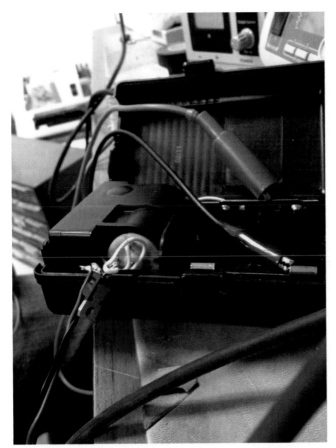

Figure 5-7. *It's a pretty simple process to wire the pump directly to the Arduino, bypassing the rocker switch and battery pack*

Build the Lego Model

Next, let's build the actual Lego chassis.

1. Let's begin with the base. Secure four Technic bricks with a 6x10 plate (see Figure 5-8).

2. Next, throw down a 2x10 plate. Figure 5-9 shows where it goes.

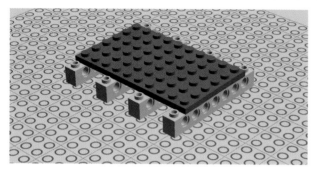

Figure 5-8. *Step 1: Securing the bricks and plate*

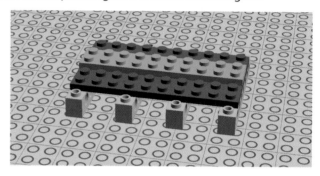

Figure 5-9. *Step 2: Adding some more beams*

3. Two more 2x10s (see Figure 5-10)!

4. Four 4x10 plates complete the square, as shown in Figure 5-11.

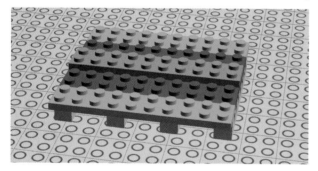

Figure 5-10. *Step 3: And another two plates*

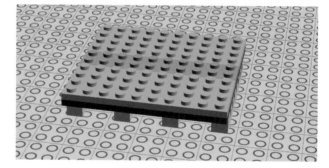

Figure 5-11. *Step 4: Adding four more plates*

5. Next, add two 1x2 bricks with cross holes and a Technic beam (see Figure 5-12).

6. Add two connector pegs with cross-axles and two standard black connector pegs as shown in Figure 5-13.

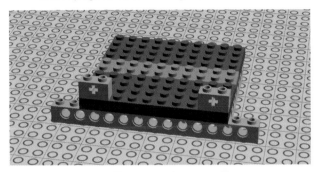

Figure 5-12. *Step 5: Adding bricks with cross holes*

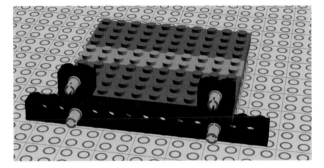

Figure 5-13. *Step 6: Inserting the pegs*

7. Technic bricks go on the other side as shown in Figure 5-14.

8. You'll need some more connector pegs (see Figure 5-15).

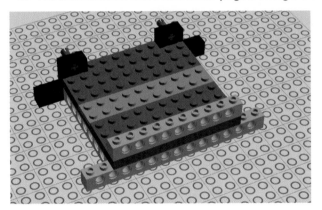

Figure 5-14. *Step 7: Connecting some bricks*

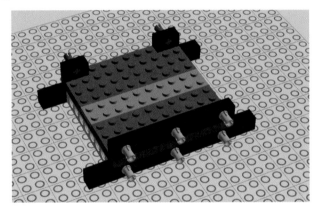

Figure 5-15. *Step 8: Adding more pegs*

9. Secure the base with these five 3M Technic beams. Figure 5-16 shows how this goes together.

10. Add two Technic bricks and two 3M pegs (see Figure 5-17).

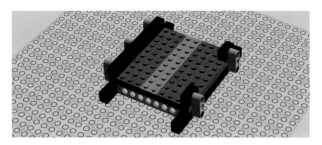

Figure 5-16. *Step 9: Securing the base*

Figure 5-17. *Step 10: Adding bricks and pegs*

11. That was so fun, let's do it on the other side as well! It should look like Figure 5-18.

12. Add bricks to each side, secured by the 3M connector pegs, as shown in Figure 5-19.

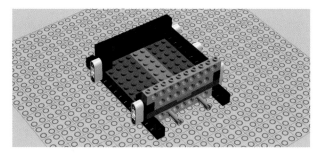

Figure 5-18. *Step 11: Duplicating the design on the other side*

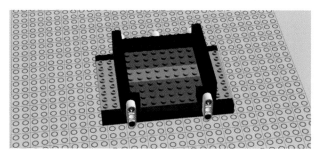

Figure 5-19. *Step 12: Adding and securing bricks*

13. Add some more 3M connector pegs, as shown in Figure 5-20.

14. And more Technic bricks (see Figure 5-21)!

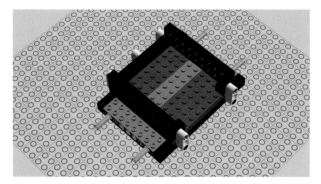

Figure 5-20. *Step 13: Adding pegs*

Figure 5-21. *Step 14: Connecting some more bricks*

Chapter 5

15. Next, let's connect four 4M Technic bricks. Figure 5-22 shows how things should appear at this point.

16. The 4x6 Technic bricks shown in Figure 5-23 finish out the base.

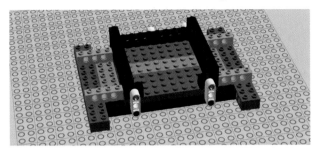

Figure 5-22. *Step 15: Attaching the 4M bricks*

Figure 5-23. *Step 16: Adding some more bricks*

17. Voila, the base is done! It should look like Figure 5-24.

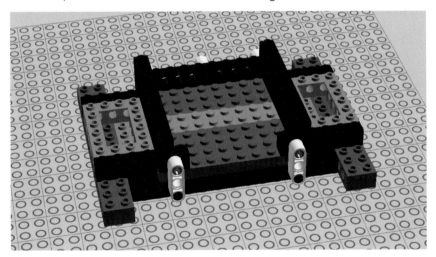

Figure 5-24. *Step 17: The base is done*

18. Next, let's build our left wall. We begin by building these vertical supports. Build eight of what you see in Figure 5-25.

19. Grab four of the supports (Figure 5-26) and set the other four aside for now.

Figure 5-25. *Step 18: Make eight of these*

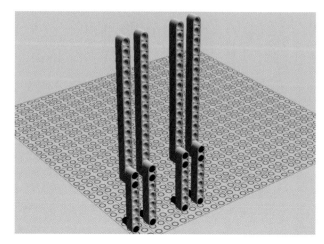

Figure 5-26. *Step 19: Four of the supports*

20. Add eight connector pegs as you see in Figure 5-27.

21. Throw some Technic beams on to secure the supports as shown in Figure 5-28.

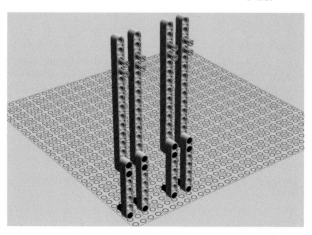

Figure 5-27. *Step 20: Adding connectors*

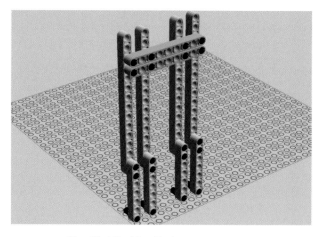

Figure 5-28. *Step 21: Attaching the beams*

Chapter 5

22. Add 2 regular and 6 cross-axle connector pegs as shown in Figure 5-29.

23. Add these halfbeam curves for added support (see Figure 5-30).

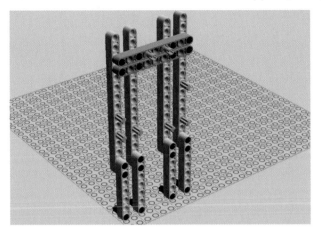

Figure 5-29. *Step 22: Adding pegs*

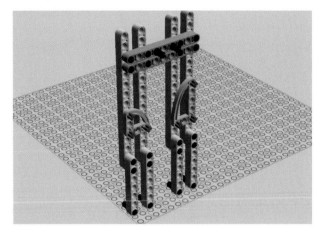

Figure 5-30. *Step 23: Attaching the halfbeam curves*

24. Next, secure them with half bushes. Figure 5-31 shows how it will look now.

25. Add more connector pegs; Figure 5-32 shows this.

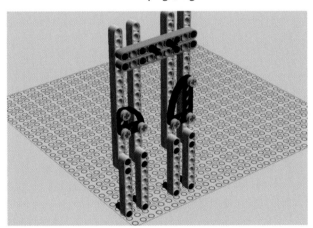

Figure 5-31. *Step 24: Securing it all with half bushes*

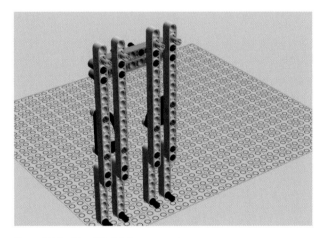

Figure 5-32. *Step 25: Adding connector pegs*

26. Place another beam, as shown in Figure 5-33.

27. You'll need to add more pegs next (Figure 5-34).

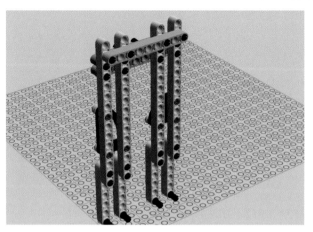

Figure 5-33. *Step 26: Installing the beam*

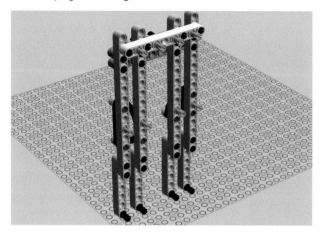

Figure 5-34. *Step 27: Attaching more pegs*

28. Add two 3M beams (Figure 5-35).

29. These angle beams will connect the left wall to the back of the robot as Figure 5-36 shows.

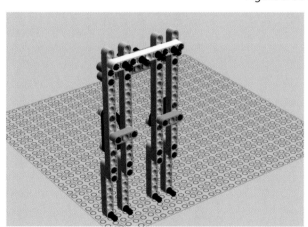

Figure 5-35. *Step 28: Attaching two beams*

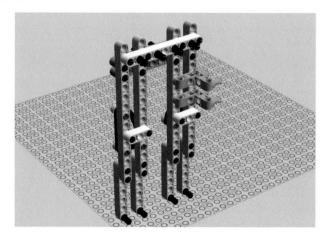

Figure 5-36. *Step 29: Adding angle beams*

Chapter 5

30. Attach this Technic brick and a 6M cross axle (see Figure 5-37).

31. Add a cross-axle extension as shown in Figure 5-38.

Figure 5-37. *Step 30: Adding a brick and a cross axle*

Figure 5-38. *Step 31: Attaching an extension*

32. The motor comes next. This will raise and lower the mixing arm. Attach it as shown in Figure 5-39.

33. Throw on a 3M beam with snaps. This will hold the activating button. Figure 5-40 shows the assembly so far.

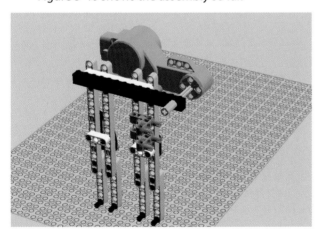

Figure 5-39. *Step 32: Snapping the motor in*

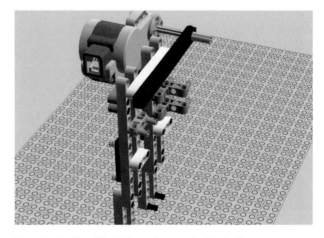

Figure 5-40. *Step 33: Adding a beam to hold a button*

34. A Mindstorms touch sensor will serve as the button that will activate the robot. Attach it as shown in Figure 5-41.

35. The left wall is done! It should look like Figure 5-42.

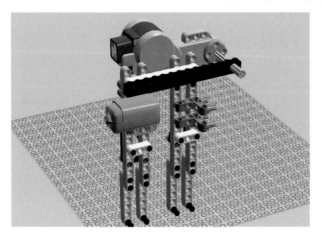

Figure 5-41. *Step 34: Attaching the button*

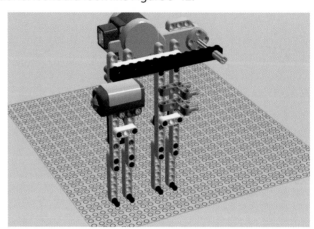

Figure 5-42. *Step 35: The finished left wall*

36. Connect the left wall to the base (see Figure 5-43).

37. Let's begin working on the mixing arm. Start with an 11M beam, a 15M beam, and a 3x5 liftarm as shown in Figure 5-44.

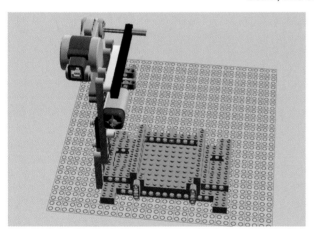

Figure 5-43. *Step 36: Attaching the wall to the base*

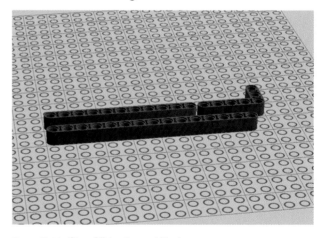

Figure 5-44. *Step 37: Laying out the beams*

Chapter 5

38. Throw on a bunch of regular and cross-axle connector pegs. Figure 5-45 shows this.

39. Add the Power Functions motor (see Figure 5-46).

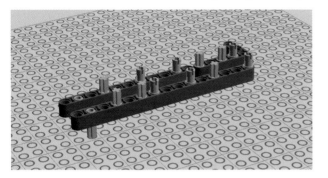

Figure 5-45. *Step 38: Adding the pegs*

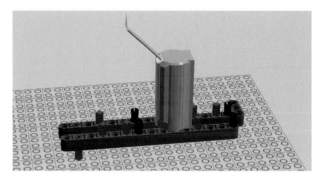

Figure 5-46. *Step 39: Attaching the motor*

40. Next come two 3M beams and 4 Technic liftarms as shown in Figure 5-47 (the liftarms are half the usual thickness, stacked).

41. Looking at the model from underneath, let's add these two catches (see Figure 5-48).

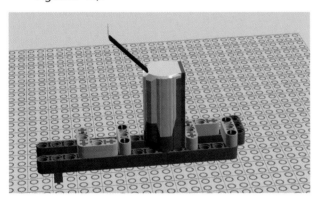

Figure 5-47. *Step 40: Adding the beams*

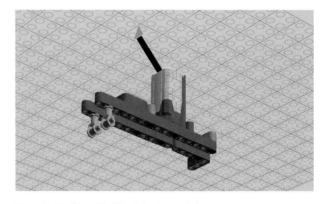

Figure 5-48. *Step 41: Attaching two catches*

42. Stick a 7M cross axle into the motor's hub. Figure 5-49 shows this.

43. Add a bush and a connector (Figure 5-50). You'll be attaching the spoon to the connector.

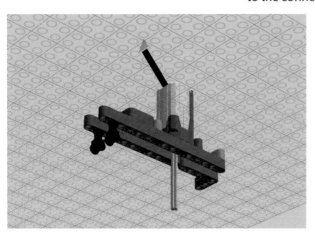

Figure 5-49. *Step 42: Inserting the cross axle*

Figure 5-50. *Step 43: Adding the bush and connector*

44. Next comes the right wall. Grab the four spare supports, shown in Figure 5-51, that you built earlier.

45. These connector pegs with cross axles come next (see Figure 5-52).

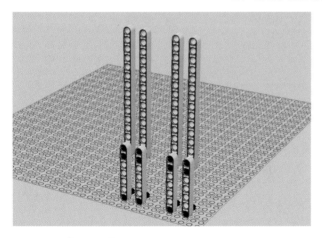

Figure 5-51. *Step 44: The four supports*

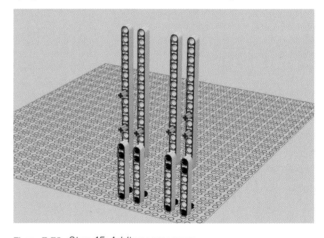

Figure 5-52. *Step 45: Adding some pegs*

46. Halfbeam curves add structural integrity. Attach them as shown in Figure 5-53.

47. And half bushes keep them in place. Figure 5-54 shows them in place.

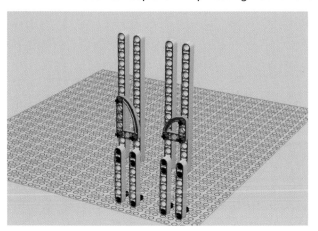

Figure 5-53. *Step 46: Adding the halfbeam curves*

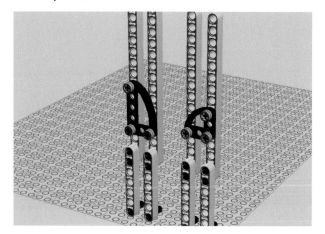

Figure 5-54. *Step 47: Securing the halfbeam curves*

48. Connect four pegs (see Figure 5-55).

49. Add a beam for a little more stability as shown in Figure 5-56.

Figure 5-55. *Step 48: Adding pegs*

Figure 5-56. *Step 49: Adding a beam*

50. More pegs! Figure 5-57 shows where they go.

51. Add two 3M beams and a big ol' Technic brick (Figure 5-58).

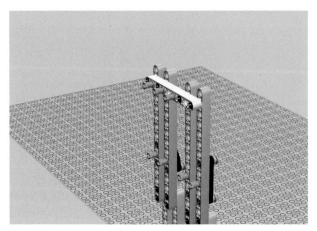

Figure 5-57. *Step 50: Inserting some pegs*

Figure 5-58. *Step 51: Attaching a brick and two beams*

52. Time for even more connectors (Figure 5-59).

53. Another touch sensor goes as indicated in Figure 5-60; it tells the Arduino when the mixing arm and syrup dispenser are down.

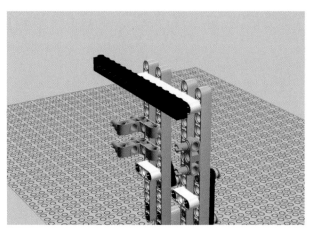

Figure 5-59. *Step 52: Adding more connectors*

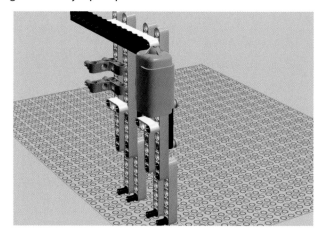

Figure 5-60. *Step 53: Attaching the motor*

54. Insert a 3M cross axle as shown in Figure 5-61.

55. A bush and a cross-axle extension are needed next. Figure 5-62 shows where you should put them.

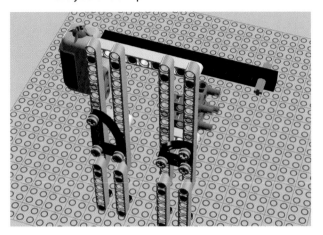

Figure 5-61. *Step 54: Inserting the cross axle*

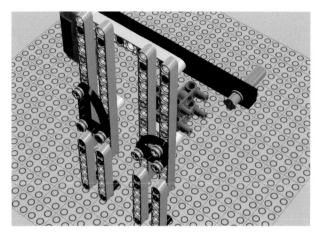

Figure 5-62. *Step 55: Adding a bush and cross-axle extension*

56. Next, add a 10M cross axle as shown in Figure 5-63.

57. Add the mixing arm to the cross axle (see Figure 5-64).

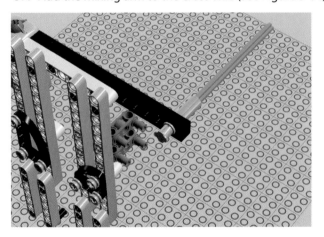

Figure 5-63. *Step 56: Attaching the cross axle*

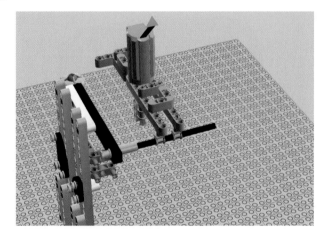

Figure 5-64. *Step 57: Connecting the mixing arm*

58. The right wall is done! It should look like Figure 5-65.

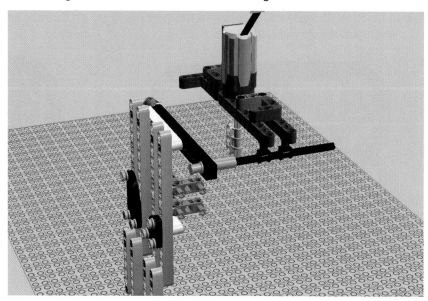

Figure 5-65. *Step 58: The completed right wall*

59. Connect the right wall and mixing arm to the left wall and base as shown in Figure 5-66.

60. Wow, that looks great! At least it does if it looks like Figure 5-67! Let's keep going.

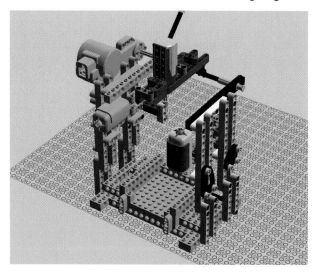

Figure 5-66. *Step 59: Combining the parts so far*

Figure 5-67. *Step 60: Admire your work*

Chapter 5

61. Let's begin working on the back. Stack four 8M and two 16M Technic bricks, as you see in Figure 5-68.

62. Connect these bricks to the angle beams (Figure 5-69).

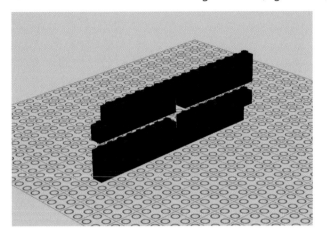

Figure 5-68. *Step 61: Prepping the beams*

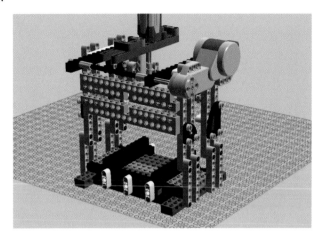

Figure 5-69. *Step 62: Attaching the beams*

63. Add some pegs as shown in Figure 5-70.

64. Now you'll add some 7M beams for support as shown in Figure 5-71.

Figure 5-70. *Step 63: Adding some pegs*

Figure 5-71. *Step 64: Adding support beams*

65. Add two cross axle pegs. These will hold the syrup dispenser (see Figure 5-72).

66. Connect four pegs to a Mindstorms motor. Figure 5-73 shows where they go.

Figure 5-72. *Step 65: Installing the pegs*

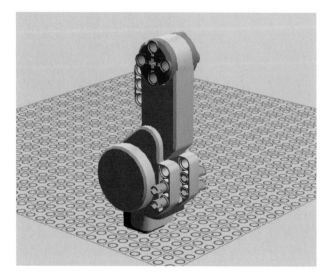

Figure 5-73. *Step 66: Attaching pegs to a motor*

67. Add two cross blocks as shown in Figure 5-74.

68. Connect the motor to the pegs you added earlier (Figure 5-75).

Figure 5-74. *Step 67: Adding cross blocks to the motor*

Figure 5-75. *Step 68: Mounting the motor*

69. Add four pegs (see Figure 5-76).

70. A pair of 7M beams and two more connectors come next (see Figure 5-77).

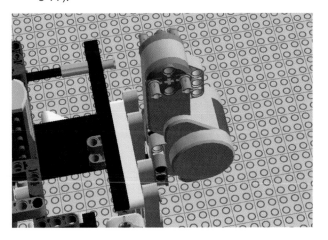

Figure 5-76. *Step 69: Placing some more pegs*

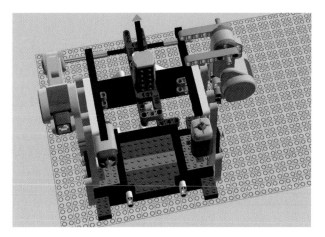

Figure 5-77. *Step 70: Adding beams and connectors*

71. Begin with a beam, then add two pegs and two 3M pegs. Figure 5-78 shows how to lay them out.

72. Add a 2x4 angle beam and a 3M beam as shown in Figure 5-79.

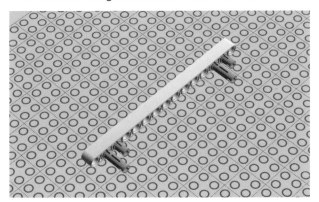

Figure 5-78. *Step 71: Preparing the beams and pegs*

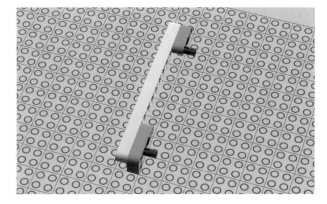

Figure 5-79. *Step 72: Adding the angle beam and 3M beam*

73. And another peg as shown in Figure 5-80.

74. Add a beam for support (Figure 5-81).

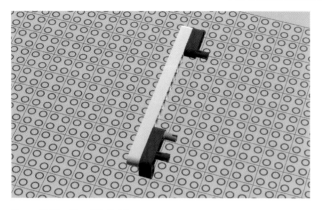

Figure 5-80. *Step 73: Adding a peg*

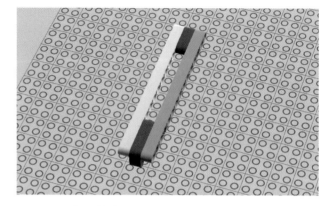

Figure 5-81. *Step 74: Connecting a beam*

75. Insert a 9M cross axle and secure with a couple of bushes. Figure 5-82 shows the assembly so far.

76. Add 8 more connector pegs (Figure 5-83).

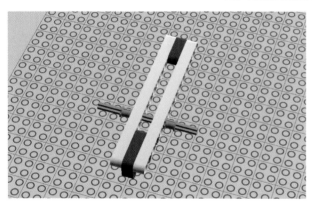

Figure 5-82. *Step 75: Securing an axle*

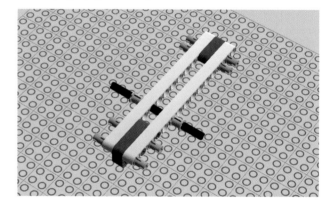

Figure 5-83. *Step 76: Adding pegs*

Chapter 5

77. Next, connect four 3x5 angle beams as Figure 5-84 shows.

78. Add a 12M cross axle and two bushes. Figure 5-85 shows where they go.

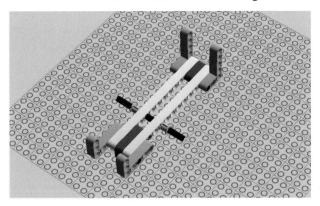

Figure 5-84. *Step 77: Connecting the beams*

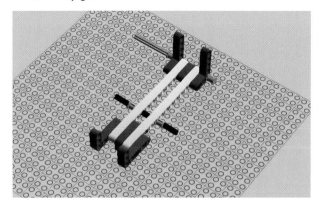

Figure 5-85. *Step 78: Adding cross axle and bushes*

79. Now you need two tubes and a bush (Figure 5-86).

80. Add two 3M beams and two bushes as shown in Figure 5-87.

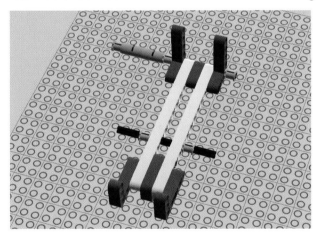

Figure 5-86. *Step 79: Adding two tubes and a bush*

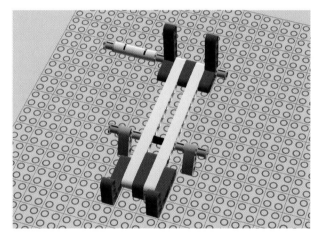

Figure 5-87. *Step 80: Securing two beams*

81. Add the syrup arm to the motor (see Figure 5-88).

82. This 4x4 angle beam, secured with a 7M cross axle and two bushes, presses on the syrup bottle to help dispense (see Figure 5-89).

Figure 5-88. *Step 81: Attaching the syrup arm*

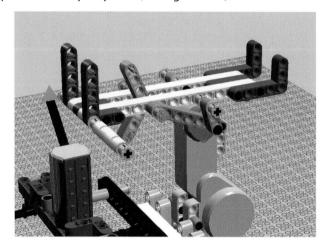

Figure 5-89. *Step 82: Attaching the beam and cross axle*

83. It is done! Your build should look like Figure 5-90.

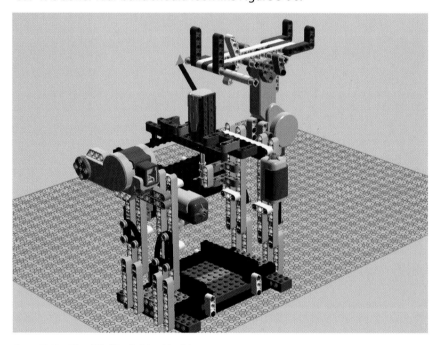

Figure 5-90. *Step 83: The finished build*

Chapter 5

Attach the Syrup Bottle

Next, let's connect the syrup bottle to the Lego chassis. As one might expect, it's actually something of a hassle to connect non-Lego to Lego—we chose to use zip ties (Figure 5-91) because of their flexibility and ease of replacement.

We put one tie at the bottom of the bottle because its soft plastic was the most rigid there; the other tie went around the bottle's nozzle. We left the middle alone, not only because the plastic was super soft and flexible there, but because that's what we need to press on in order to dispense the syrup.

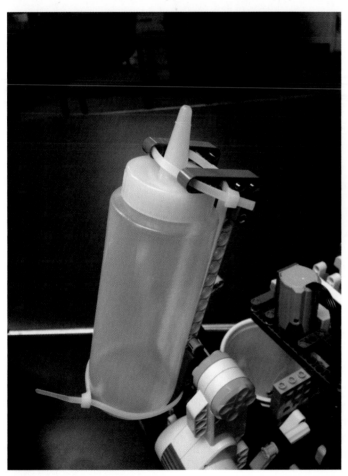

Figure 5-91. *Secure the bottle with zip ties*

Build the Mixing Attachment

Lacking any sort of obvious Lego mixing attachment, we built our own (Figure 5-92). We used a plastic spoon, cut off the handle, and then screwed it into a Lego connector. See Step 43 of the Lego build to see where you attach the spoon.

Figure 5-92. *Drill a couple of holes in your spoon and attach it to the mixer arm!*

Wire up and Install the Electronics

If you use the Bricktronics Shield, wiring up the Chocolate Milk Maker is a cinch! Simply connect the shield to your Arduino and connect the Arduino to the robot with the help of a mounting plate as you did in previous projects (see Figure 5-93). This project is a little more complicated because we use a Power Functions motor, which uses a proprietary connector. Never fear, however—we'll show you how to perform every step!

84. Attach the Arduino to a mounting plate as you did in previous projects.

85. Connect the mounting plate to the robot. There are a number of different parts of the robot you could use, and we found advantages and disadvantages to all of them. For instance, mounting the plate on the back of the robot exposes the Arduino to drips from the stir arm.

86. Mount the Bricktronics shield to the Arduino.

87. Plug in the various wires as shown in Figure 5-94.

Chapter 5

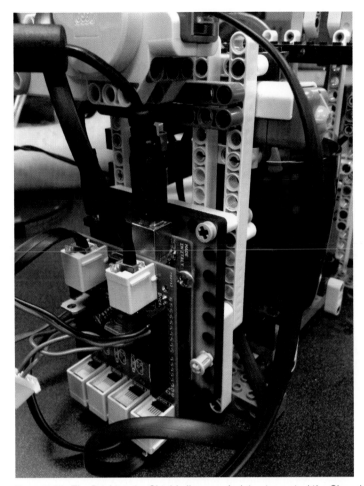

Figure 5-93. *The Bricktronics Shield allows an Arduino to control the Chocolate Milk Maker*

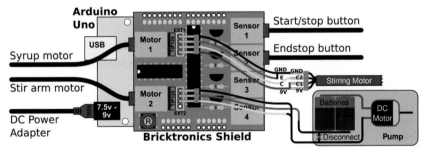

Figure 5-94. *Wire up the motors and buttons as you see here*

Adapting the Power Functions Wire

Next, we'll connect the Power Functions motor to the Arduino. Power Functions uses Lego-like connectors (Figure 5-95) so we'll need to adapt it to enable our Arduino to talk to the motor.

Figure 5-95. *Need more wire for your Power Functions motor? Get one of these inexpensive extender cables.*

Rather than damaging our motor's wire, we bought a Power Functions Extension Wire (P/N 8886) and cut it in half, wiring one end to a Molex connector (Figure 5-96). This enables us to plug our motor into the Arduino or a breadboard without difficulty.

Figure 5-96. *Attach a Molex connector, wiring it up as you see here*

You're done!

> *What if you don't have or don't want a Bricktronics Shield? In Chapter Ten we'll show you how to breadboard up the equivalent using the same components that are found in the shield.*

Chapter 5

Program the Robot

The last step—beyond drinking all that chocolate milk!—is to upload the sketch to the Arduino. As mentioned in previous projects, the code is one of the example programs included with the Bricktronics library downloadable at *http://wayneandlayne.com/bricktronics*.

```
#include <Wire.h>
#include <Adafruit_MCP23017.h>
#include <Bricktronics.h> ❶

// Make: Lego and Arduino Projects
// Chapter 5: Chocolate Milk Maker
// Website: http://www.wayneandlayne.com/bricktronics/

#define PUMP 13
#define STIR 11

#define SYRUP_WAIT 30000 ❷
#define PUMP_TIME 22000 ❸
#define STIR_TIME 20000 ❹

Bricktronics brick = Bricktronics();
PIDMotor syrup_arm = PIDMotor(&brick, 1); ❺
PIDMotor stir_arm = PIDMotor(&brick, 2); ❻
Button startstop = Button(&brick, 1); ❼
Button endstop = Button(&brick, 2); ❽

void setup() ❾
{
  Serial.begin(115200);
  Serial.println("starting!");
  brick.begin();
  syrup_arm.begin();
  stir_arm.begin();
  startstop.begin();
  endstop.begin();
  setup_loads();
}

void setup_loads() ❿
{
  pinMode(PUMP, OUTPUT);
  digitalWrite(PUMP, LOW);

  pinMode(STIR, OUTPUT);
  digitalWrite(STIR, LOW);
}

void start_pump() ⓫
{
  digitalWrite(PUMP, HIGH);
}

void stop_pump() ⓬
{
  digitalWrite(PUMP, LOW);
}

void start_stir() ⓭
```

```
  {
    digitalWrite(STIR, HIGH);
  }

  void start_stir(int speed) ⓮
  {
    analogWrite(STIR, speed);
  }

  void stop_stir() ⓯
  {
    digitalWrite(STIR, LOW);
  }

  void loop() ⓰
  {
    Serial.println("Starting loop!");

    wait_for_start_press_and_release();

    if (endstop.is_pressed())
    {
      Serial.println("Error. Endstop is already pressed at start of run.");
      return;
    }

    pump_milk();
    dispense_syrup();
    drop_stir_arm();
    stir_chocolate_milk();
    raise_stir_arm();
  }

  void wait_for_start_press_and_release() ⓱
  {
    Serial.println("Waiting for start press.");
    while (!startstop.is_pressed()) {
      //wait for start to be pressed
    };
    delay(50); //debounce
    while (startstop.is_pressed()) {
      //wait for start to be released
    };
    delay(50); //debounce
    Serial.println("Start button released!");
  }

  void pump_milk() ⓲
  {
    Serial.println("Starting pump.");
    start_pump();

    unsigned long end_time = millis() + PUMP_TIME;
    while (millis() < end_time) {
      if (startstop.is_pressed())
      {
        Serial.println("Pump stopped due to button press.");
        break;
      }
      delay(50);
```

```
  }
  stop_pump();
}

void dispense_syrup() ⑲
{
  Serial.println("Advancing syrup arm until endstop.");
  syrup_arm.set_speed(255);
  while (!endstop.is_pressed()) {
    //do nothing ⑳
  };
  syrup_arm.encoder->write(0); ㉑
  syrup_arm.stop();
  Serial.println("Endstop pressed!");
  Serial.println("Waiting and dispensing syrup.");

  for (int i = 0; i < 40; i++) ㉒
  {
    syrup_arm.go_to_pos(-100);
    Bricktronics::delay_update(SYRUP_WAIT/80, &syrup_arm); ㉓
    syrup_arm.go_to_pos(0);
    Bricktronics::delay_update(SYRUP_WAIT/80, &syrup_arm);
  }

  Serial.println("Retreating syrup arm!"); ㉔
  syrup_arm.go_to_pos(20);
  Bricktronics::delay_update(100, &syrup_arm);
  syrup_arm.go_to_pos(50);
  Bricktronics::delay_update(100, &syrup_arm);
  syrup_arm.go_to_pos(225);
  Bricktronics::delay_update(1000, &syrup_arm);
  syrup_arm.go_to_pos(350);
  Bricktronics::delay_update(1000, &syrup_arm);
  syrup_arm.stop();
}

void drop_stir_arm() ㉕
{
  Serial.println("Advancing stir arm until endstop.");
  stir_arm.set_speed(-100);

  while (!endstop.is_pressed()) {
    //do nothing
  };

  stir_arm.encoder->write(0);
  stir_arm.stop();
  Serial.println("Endstop pressed!");
}

void stir_chocolate_milk() ㉖
{
  Serial.println("Starting to stir");
  start_stir(255);
  unsigned long end_time = millis() + STIR_TIME;
  while (millis() < end_time) {
    if (startstop.is_pressed())
    {
      Serial.println("Stir stopped due to button press.");
      break;
```

```
      }
      delay(50);
    }
    stop_stir();
}

void raise_stir_arm() ㉗
{
    Serial.println("Retreating stir arm!");

    stir_arm.go_to_pos(-35);
    Bricktronics::delay_update(1000, &stir_arm);
    stir_arm.go_to_pos(-60);
    Bricktronics::delay_update(2000, &stir_arm);
    stir_arm.go_to_pos(-100);
    start_stir(85);
    Bricktronics::delay_update(2000, &stir_arm);
    stop_stir();

    stir_arm.go_to_pos(-110);
    Bricktronics::delay_update(2000, &stir_arm);
    stir_arm.go_to_pos(-250);
    Bricktronics::delay_update(1000, &stir_arm);
    stir_arm.stop();
}
```

❶ These three lines let the Arduino sketch use the Bricktronics library code that simplifies working with motors and sensors.

❷ SYRUP_WAIT is how long, in milliseconds, to wait while dispensing syrup.

❸ PUMP_TIME is how long, in milliseconds, to wait for the pump to pump milk.

❹ STIR_TIME is how long, in milliseconds, to stir.

❺ The syrup_arm PIDMotor object corresponds to the stir arm motor plugged into Motor Port 1.

❻ The stir_arm PIDMotor object corresponds to the syrup arm motor plugged into Motor Port 2.

❼ The startstop Button object corresponds to the start/stop button in Sensor Port 1.

❽ The endstop Button object corresponds to the endstop button in Sensor Port 2.

❾ The setup() function is called only once right when the Arduino is powered on. It sets up the serial port, prints a startup message, and then initializes the hardware and the software.

❿ setup_loads() initializes the pins used to start the pump and the stirring Power Functions motor.

⓫ start_pump() is a simple function that starts the pump.

⓬ Similarly, stop_pump() stops the pump.

⓭ start_stir() starts the stirring motor at full speed.

Chapter 5

⑭ This is a variant of the start_stir() function that takes a parameter that determines how fast to spin the stirrer.

⑮ stop_stir() stops the stirrer.

⑯ After setup() finishes, loop() runs over and over.

⑰ wait_for_start_press_and_release() just waits until the start button is pressed and then released. After that, it returns.

⑱ pump_milk() starts the pump, and waits for PUMP_TIME milliseconds to stop pumping. While it waits, it watches the start/stop button for a press, indicating emergency stop!

⑲ When dispense_syrup() starts, the system doesn't know exactly where the syrup arm is, just that it's vaguely upright. This function advances the arm until it hits the endstop, then it marks that as position 0. Then it pushes into the stop, repeatedly, to dispense syrup. It stops after SYRUP_WAIT milliseconds, at which point it slowly moves back to the upright position.

⑳ Wait here until the endstop is pressed.

㉑ This sets the current position of the motor to 0.

㉒ At this point, the syrup container is pointed into the cup. To help the syrup drip out, we push the syrup container and then release it, 40 times. The length of each push is calculated so we are dripping for a total of SYRUP_WAIT milliseconds.

㉓ This is a shortcut for a common idiom with the PIDMotor object. For example, Bricktronics::delay_update(1000, &motor); runs motor.update() for 1000 milliseconds before returning.

㉔ This multistep movement of the arm is supposed to start slowly, then finish, to help prevent the syrup from flying out the end when the arm is pulled back.

㉕ drop_stir_arm() slowly drops the stir arm until it presses the endstop. Then it marks that position at zero.

㉖ stir_chocolate_milk() is very similar to pump_milk(). It starts to stir, then waits for STIR_TIME milliseconds before stopping the stirrer. While waiting, it watches for start/stop to be pressed, which would indicate an emergency stop.

㉗ raise_stir_arm() tries to prevent a mess. It does this by slowly raising the stir arm a little bit, so the spoon is mostly out of the milk, but it's still over the cup. It then slowly rotates the spoon, trying to get all the milk off. Then it raises the arm all the way.

This project can be extended or tweaked in many ways! Two ideas are adding an ultrasonic sensor to detect the milk level and provide feedback, and adding a bell that the stir arm hits so you can hear when your milk is ready. Have fun thinking of more over a nice glass of chocolate milk.

PROPORTIONAL-INTEGRAL-DERIVATIVE (PID) CONTROL

When trying to drive a DC motor with precision, the first thing many people try is simply controlling how long the motor is on. Maybe they do some experiments and find that they can move their load from one side of the model to the other when they drive it for 300ms. Then they need to move it back, so they try to drive the motor backwards for another 300 ms, but it doesn't end up even close to where they started! Of course, depending on how you have everything configured, this may work fine once or twice, but given the gearing and other properties of the Lego motors, you can achieve much better and more reliable control using feedback.

Feedback is when the output of a process is "fed back" in some form into the input, to help adjust how you drive the motor. The Lego motors have internal encoders that tell the microcontroller the speed and relative location of the motor shaft. If we need to move the motor to a specific location, we can use the encoder information to change the speed we're driving the motor at.

A very common type of feedback is proportional-integral-derivative, or *PID* (Figure 5-97). For our motor control, we have a target angle or set point, which is our desired angle for the motor shaft. One example would be "rotate the motor shaft to the 2 o'clock position." To get there, we change the speed and direction of the motor, repeatedly, until we've reached the set point. The difference between our current angle and the set point is the *current error*, and our goal is to minimize the current error. We get our motor speed and direction by adding three terms: the *proportional* term, the *integral* term, and the *derivative* term.

The proportional term is related to the current error. It's found by taking the current error and multiplying it by a constant. This term makes the system go fast towards the set point when it's far away, and more slowly towards the set point when it's getting close.

The integral term is related to all the past errors, and is found by taking the sum of all the previous errors, and multiplying that by a constant. The integral term takes time into account—the more time the system spends away from the set point, the larger the integral term becomes. For example, if your motor is driving an arm that gets snagged and the proportional term isn't large enough to overcome the obstacle, the integral term will grow with each reading and increase the motor speed. The proportional term doesn't change unless the error changes, and if the arm is stuck; the error stays the same. The integral term helps us get unstuck and it can also increase the speed at which we reach the set point.

The derivative term is related to future errors—it's found by taking the difference between the current error and the last error, and multiplying that by a constant. This is like letting up on the brakes a little before coming to a complete stop. It helps prevent overshoot.

Each of the three terms has an associated constant. Finding the appropriate constants is called *PID tuning*. There are some general heuristics (methods) to help find good constants, as well as very complicated tuning methods and expensive software. Sometimes a constant will be set to zero. This eliminates that term in the motor speed calculation. This is often done with the derivative term, as the derivative term is very sensitive to noise.

After all of this, if we've picked good constants, eventually the system stabilizes at the set point.

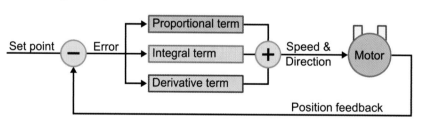

Figure 5-97. *Error correction via algorithm!*

The Next Chapter

In Chapter Six we'll jump into the world of electronics, covering everything from Ohm's Law to the myriad of sensors that could be plugged into an Arduino. It's important information that will help you understand the stuff in this book and will assist in your future projects.

Basic Electronic Theory

§

Figure 8-1. *What can you do with these obscure components? This chapter will give you the scoop*

The reason that Mindstorms and Arduino can work so well together is that both systems are based on standard electronics components and concepts. In this chapter, you will learn all about basic electronic theory in a practical, project-oriented way. We start with a quick introduction to the basic electronic components like resistors and capacitors. After that we'll introduce a variety of interesting sensors you might want to use in your own Lego and Arduino projects.

Basic Concepts in Electricity

Let's learn a bit about some basic electricity concepts such as voltage and current. Electricity is the physical phenomenon of moving electric charges (electrons), but for most of the things we'll do, it's simply the flow of electrons through metal wires and circuit components. The three most important properties of the flow of electricity in a circuit are the voltage, the current, and resistance.

Voltage, Current & Resistance

Voltage

Sometimes called "electric potential," this is a measurement of how much the electrons want to move from one location to another. The way it works is that the higher the voltage, the stronger the "pressure" for the electrons to move from one location to the other. Voltage is measured between two points, such as between the + and - terminals of a battery, or between two points in a circuit. Voltage is measured with "Volts" and is abbreviated with the letter "V."

Current

Abbreviated with the letter "I," this is the flow of electric charge, and is measured with units of "Amperes." The higher the current, the greater the flow of electricity through the component. If too much current flows through a circuit, it can cause it to heat up and cause damage, so there are many devices (such as a fuse, circuit breaker, or resistor) that are used to limit currents.

Resistance

This is the third important concept of basic electronics. It describes to what degree circuit components (even wires!) will resist the flow of electric current. The incandescent light bulbs in your house resist the flow of the electricity and produce heat and light. The core idea of resistance is that the voltage *across* a resistor is directly related to the current flowing *through* the resistor. The relationship between voltage and current of a resistor is a simple linear relationship based on the resistance value of the resistor, and is called *Ohm's Law*.

Ohm's Law

Ohm's Law relates the voltage *across* a resistor with the current *through* the resistor, based on the resistance value. It can be written in three equivalent ways (* is used as the multiplication symbol):

```
V = I * R
R = V / I
I = V / R
```

To help remember Ohm's Law, many people like to visualize it as a triangle, as seen in Figure 6-2.

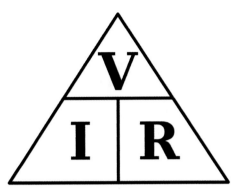

Figure 6-2. *Ohm's Law is an easy-to-understand formula that is the basis for much of hobbyist electronics*

To read this triangle diagram, first look at the value you want to solve for, then look at the other two variables to see their relationship. For example, to solve for the current I, we see that V is positioned above R, just like it is in the fraction V / R, telling us that the current is equal to the voltage divided by the resistance, or I = V / R. Similarly for the voltage V, the variables I and R are adjacent, just like in the product I * R, telling us that V = I * R. Pretty simple, right?

If you know at least two of the three important values about a resistor (voltage across, current through, or the resistance) then you can use Ohm's Law to find the missing value.

> *When you use Ohm's law, resistance (R) is measured in ohms, voltage (V) is measured in volts, and current (I) is measured in amps.*

For example, if you connect a 1 kilohm (1,000 ohm) resistor to a 5-volt power supply, how much current will it draw? We want to solve for current, so we use the equation I = V / R. Filling in the values for voltage and resistance, we find that I = 5 / 1000 = 0.005 amps, or 5 milliamps (mA).

Other Concepts

While V, I, and R are the core principles around which electronic theory is focused, there are other important terms to learn.

Ground

Ground (Figure 6-3) usually means the reference voltage in your circuit, and the return path for electric current. If your circuit is battery powered, then "ground" most likely means the voltage at the negative terminal of your battery. If your circuit is powered from an AC-DC voltage supply (wall wart, etc.) then "ground" most likely means the negative terminal of the power supply. If your circuit is powered from AC (like household appliances) then "ground" usually means the "earth" wire (either bare or with green insulation) that is connected to the appliance chassis.

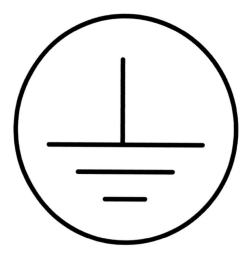

Figure 6-3. *The symbol for ground*

Series & Parallel Connections

Two components are *connected in series* if they are connected in a line (sometimes called *daisy-chained*). In a series-connected circuit, the same amount of current flows through all components, while each component might have a different voltage across it. A disadvantage to series-connected components is that if one fails, the whole path is affected. This can be seen in very old strings of lights, where if one bulb burns out, the whole strand of lights goes dark.

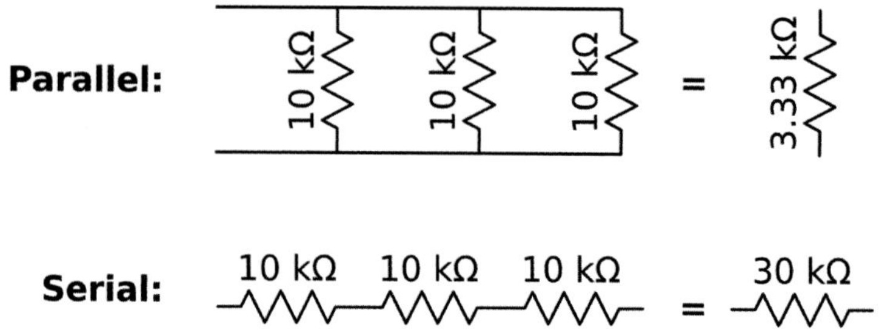

Parallel: 10 kΩ 10 kΩ 10 kΩ = 3.33 kΩ

Serial: 10 kΩ 10 kΩ 10 kΩ = 30 kΩ

Figure 6-4. *The difference between series and parallel connections*

If you connect some resistors together in a series connection, the total resistance from one end to the other is simply the sum of the individual resistances:

R = R1 + R2 + ... + Rn

For example, if you had three 10k ohm resistors connected in series, their equivalent resistance would simply be their sum, or 30k ohms of resistance.

Two components are *connected in parallel* if both ends of all connected components are connected together, shown in Figure 6-4. The current in a parallel-connected circuit is divided (not necessarily evenly) between all components. However, all components have the same voltage.

If you connect some resistors together in a parallel connection, they will evenly divide the total current, but finding the total resistance is more complex than for the series case. The equivalent resistance is the following:

```
R = 1 / ( 1/R1 + 1/R2 + ... + 1/Rn )
```

For example, if you had three 10k ohm resistors connected in parallel, their equivalent resistance would be 1 / (1/10k + 1/10k + 1/10k) = 3333.33 ohms.

TOOL OF THE TRADE: THE MULTIMETER

A multimeter (Figure 6-5) is the most common tool used for simple electronics. It's a device capable of accurately measuring various properties of a circuit, including but not limited to voltage, current, and resistance. Every model of meter is slightly different, but here are some commonplace properties that can be measured.

Voltage

Voltage is always measured *between* two points. First, put the multimeter into a voltage-measuring mode, commonly labeled "VDC." Connect the two probes to the two points in your circuit to measure the voltage between those two points. If someone asks you for the voltage at a single point in your circuit, they are probably asking for the voltage between that point and the circuit's ground (see "ground," earlier in this chapter). When your multimeter is set to voltage measuring mode, the resistance between the two probes is very large, so that the meter barely affects the voltage at all.

Current

Current is always measured *through* part of your circuit, meaning that you must disconnect a wire and place your meter *in the path* of the current you want to measure. For example, if you want to measure the current through an LED and its resistor, you need to disconnect that circuit path, and insert the multimeter into the current path you want to measure. When your multimeter is set to current measuring mode, the resistance between the two probes is very very small, so that the meter barely affects the current at all. When measuring current, it's important to never measure the current across the terminals of a voltage supply, like a battery or DC adapter. Measuring the current through a circuit using a voltage supply is fine, but connecting the meter across the terminals of a voltage supply is effectively shorting the supply!

Resistance

Measuring the resistance of a resistor is very easy to do with your multimeter. Simply set the meter to resistance mode and place the two leads on opposite sides of the resistor. This can be very useful if you have trouble reading the colored stripes on most resistors, or have some resistors without colored stripes. If your resistor is still connected to a circuit, measuring the resistance this way probably won't work, because the circuit will affect the measurement.

Continuity

Most multimeters also have a very useful setting where it will make an audible BEEP! when the two leads are electrically connected. This makes it really easy to investigate new devices, such as switches, wiring harnesses, sockets and plugs, etc., to determine which wires are connected together.

Temperature

Some meters have a passive infrared heat sensor, and possibly a physical temperature probe as well. This is useful for taking precise measurements in degrees, with both Fahrenheit and Celsius typically supported.

There are many more factors that specialty meters measure, such as sound, inductance, frequency, and conductance, but the items above are the big ones. If you want to learn more about multimeters and how to use them, we suggest Ladyada's Multimeter Tutorial: *http://www.ladyada.net/learn/multimeter/*

Figure 6-5. *The multimeter is one of the most important tools an electronics hobbyist—or professional!—can own*

Know Your Electronic Components

Next, we'll describe some of the more common components that you'll use in your projects. Chances are, if you've built any of the projects in the book you'll have at least a passing familiarity with these parts, but if not, read on!

Resistors

A resistor (Figure 6-6) is the simplest of all circuit components; it simply resists the flow of electricity. A resistor has two terminals and is not polarized. Resistors have many uses and applications in a circuit, including limiting the flow of electricity to protect delicate components, making filters to remove unwanted signal noise, providing default voltage values for switches (as a pull-up or pull-down resistor), reducing voltages with a voltage divider, etc.

Figure 6-6. *A resistor*

Determining a Resistor's Rating

Common resistors are marked with a series of colored bands (Figure 6-7) to indicate their resistance value. Some resistors have five bands of color, but most have only four, but you read their values off their color bands in almost exactly the same way.

One of the colors on the end will be gold, silver, or brown, indicating the *tolerance* of the resistor. This tolerance indicates how precise the resistor's value is, with gold being 5%, silver 10%, and brown 1%. Rotate the resistor so the tolerance stripe (usually shiny gold or silver) is on the right side, and then read the colors from left to right. You'll use the color of the first two bands (or the first three bands on a 5-band resistor) to look up the value in the diagram below, and simply combine them. For example, if the first color is yellow (4) and the second is violet (7), then you start with the number "47." Then, the last band before the tolerance band is used to look-up the *multiplier*, which is a power of 10. For example, if the multiplier color is red, that corresponds to a multiplier of 100x. To find the resistor's value, take the original value (47 in this example) and multiply it by the multiplier from the next-to-last band (100x in this example). This value is 4700, usually printed as 4.7k.

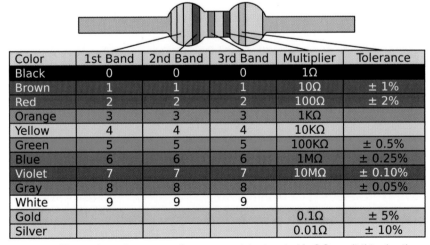

Color	1st Band	2nd Band	3rd Band	Multiplier	Tolerance
Black	0	0	0	1Ω	
Brown	1	1	1	10Ω	± 1%
Red	2	2	2	100Ω	± 2%
Orange	3	3	3	1KΩ	
Yellow	4	4	4	10KΩ	
Green	5	5	5	100KΩ	± 0.5%
Blue	6	6	6	1MΩ	± 0.25%
Violet	7	7	7	10MΩ	± 0.10%
Gray	8	8	8		± 0.05%
White	9	9	9		
Gold				0.1Ω	± 5%
Silver				0.01Ω	± 10%

Figure 6-7. *Want to know how many ohms your resistor is rated for? Consult this chart!*

Capacitors

A capacitor (Figure 6-8) is a device used to temporarily store energy in an electric field. It is similar to a battery, but uses two parallel pieces of metal that are very close but not touching (like the capacitor schematic symbol) to store electric charge, instead of storing electric charge in chemical energy like a battery. Capacitors are also smaller, more lightweight, and can charge and discharge much faster than batteries. Capacitors have many uses in a circuit, including filtering noise from signals, helping to isolate circuit components, power supply conditioning, etc. The most common use for capacitors in hobby circuits is to help the power supply. When something quickly changes its power demand, it can have an effect on the rest of the circuit. Decoupling capacitors can help provide a quick burst of electricity to compensate for sudden increases in power demand.

Figure 6-8. *Capacitors help regulate power in a circuit, among their other uses*

Diodes and LEDs

Diodes are electrical components that only allow current to flow through them in one way. A light emitting diode, or LED (Figure 6-9), is nothing more than a fancy diode that also happens to emit visible light when the current is flowing the correct way. Diodes have two terminals and are polarized, so you must connect them the correct way. For LEDs, one leg will be longer, indicating the positive terminal. For most diodes, a gray or black band on one end of the diode indicates the negative terminal.

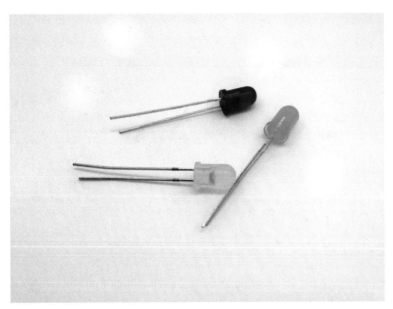

Figure 6-9. *Diodes permit current to travel in only one direction*

Inductors

An inductor (Figure 6-10) is a device to store energy in a magnetic field, usually in the form of a coil of wire, sometimes with a core of magnetic metal. It does this by using the electric current to generate a magnetic field in the coils of wire that make up the inductor, and resists changes to the current flowing through the inductor. Inductors are much less frequently used than resistors and capacitors, but can still be useful to the hobbyist. Because motors behave very similarly to inductors, sometimes we model a motor as a simple inductor in our circuit diagrams. Inductors have two terminals and are generally not polarized.

Figure 6-10. *Inductors usually appear in the form of a coil of wire surrounding a core of metal. Credit: Windell H. Oskay*

Transistors

A transistor (Figure 6-11) is a circuit component that can be used to amplify signals, and also to turn things on and off. They usually have three terminals and are polarized. They are a little more complicated to use than resistors, capacitors, and diodes, but can be very useful in a circuit.

Figure 6-11. *Transistors control the flow of electricity through a circuit*

Integrated Circuits

Integrated circuits (sometimes called chips) are a piece of specially-prepared (Figure 6-12) silicon that has tens to millions of very small transistors built into it. Using a chip instead of discrete components can greatly reduce the cost and increase the speed of your circuit. There are many chips commonly available, from very simple logic gate and amplifier chips, all the way up to the very fancy chips in your computers and cell phones. A very popular hobbyist chip is the 555-timer chip, which allows you to make all sorts of signals for circuits involving lights and sounds. Another very popular type of chip is a *microcontroller chip* that is a very simple computer, where you load special codes into it to tell it what to do and how to behave.

Figure 6-12. *Integrated circuits are entire circuits embedded in a chip of silicon*

Sensors 101

Half of robotics is sensors: it's how we get information about the outside world into the robot! The Lego Mindstorms NXT system ships with four different types of sensors, including touch, light, and ultrasonic, as well as the rotation sensors built into the motors. Using these default sensors, you can make a wide variety of interesting robotics projects. However, there are many other kinds of sensors that you can easily use with your Arduino or NXT projects!

Digital Sensors

The simple sensors we'll look at in this section can be classified into two categories: digital and analog. The first type of sensor, simple binary digital sensors, output either a very low value or a high value, traditionally called logic low and logic high, respectively. They are very easy to work with using both Arduino and NXT. The following are some of the more common types of digital sensors.

Tilt Sensor

A tilt sensor (Figure 6-13) can be used to measure if the sensor is upright or not. It has a ball inside, and it has two states. When the sensor is upright, the ball doesn't make contact. When it is on its side, the ball makes contact, and the circuit is completed. It can be hooked up to an Arduino between any digital input pin and ground. Configure the pin to have an internal pullup with `pinMode(pin, INPUT_PULLUP)`, and then read the pin with `digitalRead(pin)`. If the pin reads high, that means the sensor is upright. If the pin reads low, it means the ball has made contact because the sensor is tilted.

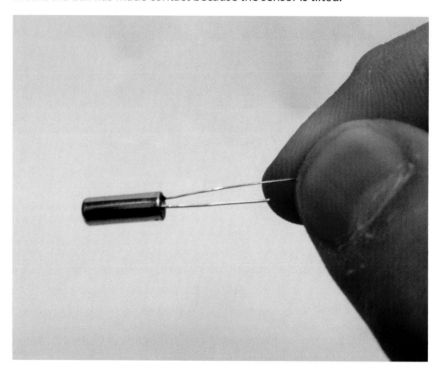

Figure 6-13. *A tilt sensor senses if your robot is upright or not*

Hall-Effect Sensor

Another easy-to-use sensor is the Hall effect sensor. This sensor is able to detect if a magnet is nearby, and even detect which pole of the magnet is nearest. The simplest kind of Hall-effect sensors (shown in Figure 6-14) have three pins: one is for power, one is for ground, and the third is the output signal. For this particular sensor, the output signal will be pulled down to 0 volts when the south pole of a magnet is detected nearby. The pin will not be pulled down to zero when no magnet is detected. By using a pullup resistor with this sensor, it will be low when the south pole of a magnet is close, and high when there isn't one close.

Figure 6-14. *The Hall effect sensor can tell if a magnet is nearby*

Passive Infrared Motion Sensor

The passive infrared (PIR) sensor, seen in Figure 6-15, detects motion due to changes in infrared radiation. All matter above absolute zero emits electromagnetic radiation due to heat, but this sensor detects the band of waves given off by warm things like humans and pets the best. The sensor is generally split into two halves, and when the IR received by one half differs from the other half, the sensor has detected movement. The easiest way to work with PIR sensors is to use a simple binary digital logic value that corresponds to either "motion detected!" or "no motion detected." You can simply connect the signal output from the sensor to a digital input pin on the Arduino, with no resistors required!

Figure 6-15. *The PIR detects infrared radiation. Credit: Adafruit Industries*

Temperature and Humidity Sensor

Some sensors are digital sensors, but output more information than just a simple on/off value. One such sensor is the DHT22 temperature and humidity sensor (Figure 6-16). It outputs a series of low and high digital values in sequence to communicate the measured temperature and humidity values. You can connect the DHT22 digital output pin to your Arduino, using a pull-up resistor between that pin and the positive power supply. The best way to use a sensor like this is to use an established software library, such as the Adafruit DHT-sensor-library: *https://github.com/adafruit/DHT-sensor-library*.

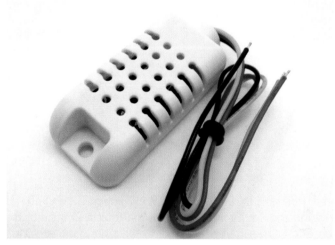

Figure 6-16. *This sensor sends back data on the temperature and humidity in the area*

Infrared Receiver

A common household infrared remote control, like you might have for your television, transmits information for each keypress using a series of infrared signals. You can easily detect and decode these IR data transmission using an infrared receiver and decoder (sometimes called a demodulator) such as the one in Figure 6-17. Most IR detectors have three leads, one each for power and ground, and the third is the decoded data output pin. When the detector receives a burst of infrared signal, it will drive the output pin to logic high, logic low if no signal is received. Your remote control will send specially-timed bursts of infrared light to transmit binary data to your television or radio, with each code representing a different operation, such as "turn off" or "raise volume." You can use an infrared detector to provide wireless data transfer capabilities to your project, and can use an infrared LED (which works just like a standard LED) to provide IR transmission capabilities.

Figure 6-17. *Your robot can detect infrared signals thanks to this IR receiver*

Analog Sensors

Many sensors are *analog sensors*, meaning that they output a range of voltages that correspond to their measured value. The Arduino has a number of special pins that can be used to capture an analog value, and you must use one of these pins when working with analog sensors.

Thermistor

A thermistor (Figure 6-18) is a very simple analog sensor. The temperature of the sensor changes its resistance. To use it with an Arduino, it can be hooked up in a voltage divider with a known resistance. Then, as the resistance varies, the voltage on the pin in the middle of the divider changes in a way we can predict. Temperature changes the resistance of a thermistor in a non-linear way, so the easiest way to get a temperature out of the measured value is to use a simple software lookup table, such as the one on the Arduino playground (*http://arduino.cc/playground/ComponentLib/Thermistor*).

Figure 6-18. *A thermistor's resistance changes as the temperature goes up*

Basic Electronic Theory

Photocell

A photocell (Figure 6-19) or photoresistor, is a sensor very similar to a thermistor, in that the amount of light striking a photocell changes its resistance. A photocell in the dark has a very high resistance, but as more and more light strikes the photocell surface, the resistance will drop, allowing you to measure the amount of light hitting the photocell. Just like for the thermistor, you should build a voltage divider with the photocell in order to detect varying levels of ambient light. You can use a photocell to build a robot that follows the brightest light it can find (like a moth!) or a robot that hides from the bright lights (like a cockroach!), among many other ideas.

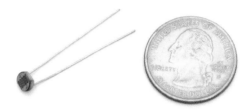

Figure 6-19. *The photocell detects light and tells the robot. Credit: Adafruit Industries*

Force Sensitive Resistor

Another simple analog sensor is the force-sensitive resistor (FSR) shown in Figure 6-20. As the FSR is bent or pressure applied to the end, its resistance will decrease in proportion to the amount of force applied. You must be very careful in how you connect wires to an FSR, as soldering to the tabs can very easily break the entire resistor!

Figure 6-20. *This resistor changes its resistance depending on how much force has been applied*

Ultrasonic Sensor

Ultrasonic sensors (Figure 6-21) use pulses of high-frequency sound waves to accurately measure the distance between the sensor and another object. One popular ultrasonic sensor is the Maxbotix LV-EZ1, which can easily output a simple analog voltage that is proportional to the measured distance. To use this kind

Chapter 6

of sensor you can simply connect the analog output signal from the sensor to an Arduino analog input pin. If there is nothing in front of the sensor (or further out than 6 meters) the Arduino will read a value above 500. A value less than 500 indicates something is approaching the sensor. The value is proportional to the distance the ultrasonic waves travel before bouncing back to the sensor.

Figure 6-21. *The ultrasonic sensor uses sound to measure distance*

Accelerometer

One of the coolest analog sensors available is the ADXL335 3-axis accelerometer, pictured in Figure 6-22. It can measure acceleration in three axes (x, y, and z) at the same time, up to 3g in each direction. Each axis has an analog output value that is proportional to the current acceleration. To use this sensor you will need one analog input pin for each axis to measure. You can connect the sensor's analog outputs directly to each analog input pin. The device outputs "ratiometric" analog values between 0 and 3.3 volts, meaning that the acceleration range of -3g to +3g is evenly spread between 0 and 3.3 volts. No acceleration (0g) is represented by approximately 1.65 volts.

Figure 6-22. *This accelerometer tells your robot whether it's moving, and how fast*

Further Study

If you want to learn more about electronics, we suggest you investigate Charles Platt's excellent MAKE: Electronics (Figure 6-23). It's packed with tutorials and teaches you the basics of every aspect of hobby electronics. Going through and completing the various projects will do much to reinforce your knowledge of the subject. The Maker Shed (*http://makershed.com*) has copies of the book, as well as component packs that supply you with the parts you need to complete every project in the book.

Chapter 6

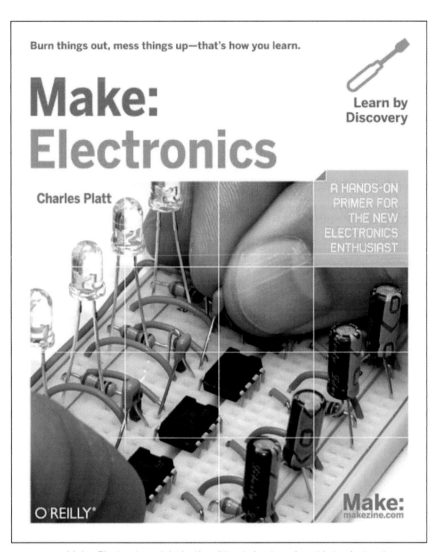

Figure 8-23. *Make: Electronics might be the ultimate beginner's guide to electronics. Check it out!*

Gripperbot | 7

Figure 7-1. *The Gripperbot is a rolling manipulator you can control with your hands' motions*

We're extremely excited about this next project. The Gripperbot is a rolling robotic arm that you control (see Figure 7-1) with a pair of gloriously nerdy arm-mounted Arduinos equipped with Wii Nunchuks and a wireless card for communicating with the robot! The roller itself packs an Arduino as well, along with a Bricktronics Shield and Bricktronics Motor Controller. It's a complex build that requires a lot of parts, but the result is so sweet!

The Gripperbot (Figure 7-2) can be reconfigured for a simpler, and more inexpensive, robot. If you remove one motor from the body, or change the control scheme, you can get by with a single bracer, which eliminates an Arduino, an XBee shield, an XBee, a Nunchuk, and a battery pack. If you remove a motor, and make sure not to spin the arm around and pull the cords out, you can get by with only the Bricktronics Motor Controller on the body, which cuts an Arduino, an XBee shield, an XBee, and a battery pack. If you skip the arm, you won't need the Bricktronics Motor Controller. Just about any single component can be removed, and it will cascade into other things you can remove—but the bot, as it is, is a fun remote-control toy that's inspired by something John's wanted since he was a little kid.

Figure 7-2. *The Gripperbot has a lot of options for customization*

Parts List

The Gripperbot is a complicated project and you'll need a lot of stuff to build it. Let's get started assembling our parts.

Electronics Parts

Figure 7-3. *In this chapter we'll get to play around with XBee wireless cards!*

- 3 Arduino Unos
- 2 XBee wireless modules (See Chapter Ten for more information on XBee modules)
- 2 Arduino Wireless Proto Shields (Maker Shed P/N MKSP13)
- 2.1 mm DC Plug (we used CP3-1000-ND from Digi-Key)
- 1 Bricktronics Shield
- 1 Bricktronics Motor Controller
- 2 Solarbotics NunChucky boards (Solarbotics P/N 31040) (http://www.solarbotics.com/product/31040/)
- 2 Wii Nunchuks
- Battery packs (we used Mouser P/N 12BH361A-6R and Jameco SBH-431-1AS-R)
- Mounting plates for Arduinos and battery packs (see Chapter Ten for more information on these plates)

THE BRICKTRONICS MOTOR CONTROLLER

In Chapter One we introduced the Bricktronics Shield; now we're going to talk about its big brother, the Bricktronics Motor Controller (Figure 7-4). It's a self-contained microcontroller (no Arduino needed!) with L293D motor control chips allowing control of up to five Mindstorms or Power Functions motors. It even sports a socket for placing an XBee wireless board. It's powerful and useful!

Figure 7-4. *The Bricktronics Motor Controller*

Lego Elements

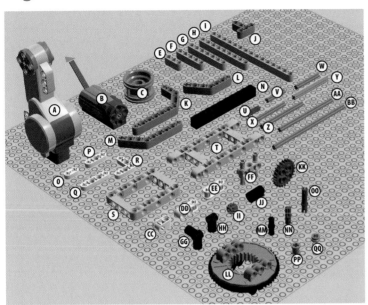

Figure 7-5. *You'll need the above Lego parts to build the Gripperbot and its control bracers*

A. 4 Mindstorms Motors

B. 2 Power Functions motors

C. 4 Rims

D. 2 Tank Treads

E. 11 3M Technic beams

F. 6 5M Technic beams

G. 12 7M Technic beams

H. 7 9M Technic beams

I. 1 11M Technic beam

J. 4 15M Technic beams

K. 6 Technic T-beams

L. 5 Technic double angle beams 3x7

M. 36 Technic angle beams 4x4

N. 12 Technic angle beams 4x6

O. 4 10M Technic bricks

P. 12 2M Technic lever beams

Q. 6 3M Technic lever beams

R. 6 4M Technic lever beams

S. 2 3x3 Technic levers ("L" shaped)

T. 3 5x7 Technic beam frames

U. 2 5x11 Technic beam frames

V. 8 2M cross axles (usually red in color)

W. 6 3M cross axles

X. 1 4M cross axles

Y. 5 5M cross axles

Z. 5 6M cross axles

AA. 10 7M cross axles

BB. 4 8M cross axles

CC. 4 9M cross axles

DD. 16 12M cross axles

EE. 20 2M cross blocks

FF. 7 3M cross blocks

GG. 9 3M double cross blocks

HH. 8 3M Technic beams with pegs

II. 4 90-degree angle elements

JJ. 2 0-degree angle elements

KK. 1 8-tooth gear

LL. 1 worm gear

MM. 1 16-tooth gear

NN. 2 24-tooth gears

OO. 2 turntables

PP. 2 tubes

QQ. 57 connector pegs w/friction tabs (usually black)

RR. 46 cross connectors (usually blue)

SS. 8 3M connector pegs (usually blue)

TT. 37 bushes

UU. 16 half bushes

Building Instructions

The Gripperbot is a complicated build, all the more so because it includes a pair of Wii controllers! Never fear, we'll show you how to build a pair of Lego arm guards (also known as bracers) as well as the robot itself.

Bracers

The first thing we're going to build is the armbands (also known as bracers) that serve as the robot's controllers (Figure 7-6). There will be two of them, one for each arm, each equipped with an Arduino, an XBee wireless module (see Chapter Ten for more details on XBees), a Solarbotics NunChucky board, and a Wii Nunchuk controller.

For each bracer, start with an Arduino Uno with an XBee module plugged into it and a Solarbotics NunChucky breakout board added on. The Nunchuk gets plugged into the NunChucky. A 3-AA battery pack gets plugged into the DC jack on the Arduino. You can see this setup illustrated in Figure 7-7.

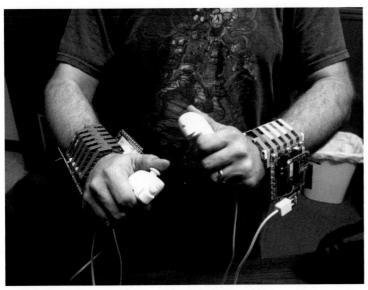

Figure 7-6. *The Gripperbot is controlled by a pair of Lego wrist guards equipped with Arduinos and wireless boards*

Wrist Assembly (2x)

AA Battery
AA Battery
AA Battery

USB jack

Arduino

Power jack

+ XBee Shield +

Solarbotics

Nunchucky

Figure 7-7. *Wiring up the bracers is super easy!*

Lastly, you can secure the bracers around your wrists by connecting the 4M Technic Lever Beams protruding from the ends, visible in Step 13 below.

1. Begin with a 12M cross axle and slide a 7M beam and a 4x6 angle beam (Figure 7-8).

2. Add one more of each (see Figure 7-9).

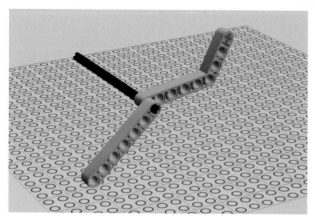

Figure 7-8. *Step 1: Lining up the beams*

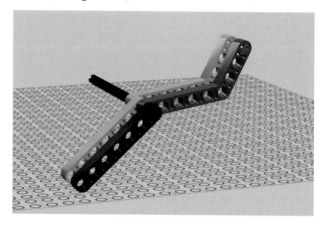

Figure 7-9. *Step 2: Adding more beams*

3. Add more of each until you max out that cross axle, as shown in Figure 7-10. Note that the 7M beams will swing freely.

4. Secure those 7M beams with another 12M cross axle (see Figure 7-11).

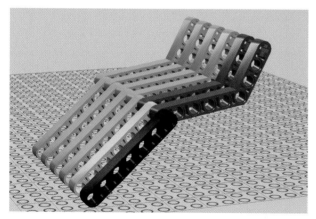

Figure 7-10. *Step 3: Filling the axle with beams*

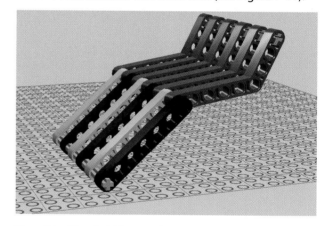

Figure 7-11. *Step 4: Adding another axle*

5. Next, grab another 12M beam and attach two 4X4 angle beams to it as you see in Figure 7-12.

6. Fill up the axle with more angle beams (see Figure 7-13).

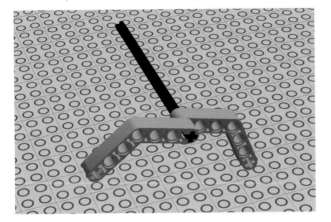

Figure 7-12. *Step 5: Attaching an axle to two angle beams*

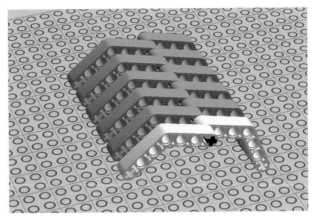

Figure 7-13. *Step 6: Fill up that axle*

7. Secure the end as you did in Step 4 (Figure 7-14).

8. Add six 3x3 angle beams to the product of steps 1 through 4 and secure with a 12M cross axle as shown in Figure 7-15.

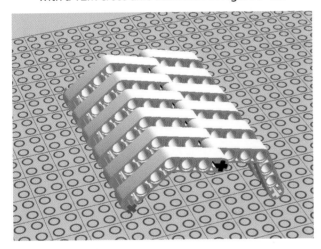

Figure 7-14. *Step 7: Securing the end*

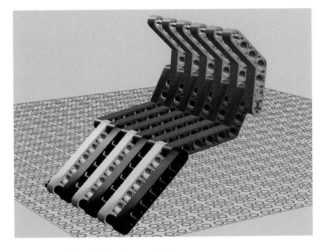

Figure 7-15. *Step 8: Adding angle beams*

9. Add the construct you built in steps 5 through 7 to the angle beams and secure with cross axle. Note that the axle goes through the smooth hole, not the cross hole, allowing it to swing freely (Figure 7-16).

10. Add ten 2M cross blocks along with two 12M cross axles (Figure 7-17).

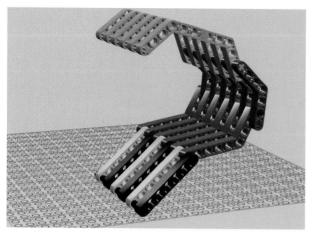

Figure 7-16. *Step 9: Combining the parts*

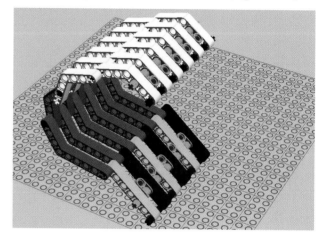

Figure 7-17. *Step 10: Adding cross blocks and axles*

11. Now, add 14 cross connectors. Note that some of them are only partially visible and the butt ends can be seen through the cross blocks in Figure 7-18.

12. Next, secure the ends of the cross axles with 2M Technic levers, as shown in Figure 7-19.

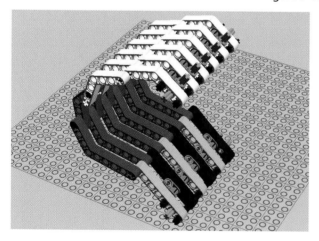

Figure 7-18. *Step 11: Adding cross connectors*

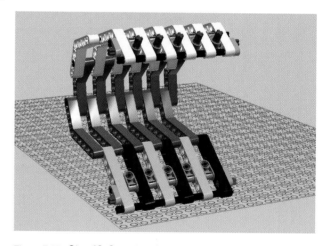

Figure 7-19. *Step 12: Securing the axle ends*

13. Add two 4M Technic levers as seen in Figure 7-20. These will keep the bracer closed when it's on your arm. You're done with this bracer! Now make another one just like it.

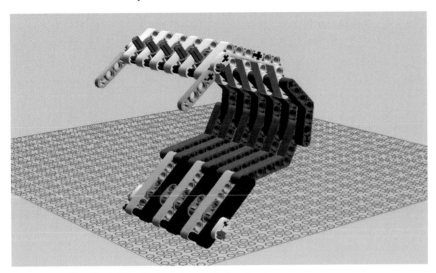

Figure 7-20. *Step 13: Adding Technic levers. You're done!*

Gripperbot

Next, let's focus on the robot itself!

14. Begin with a 5x11 Technic beam frame and add two connector pegs and four 3M connector pegs, as shown in Figure 7-21.

15. Add two beams (see Figure 7-22).

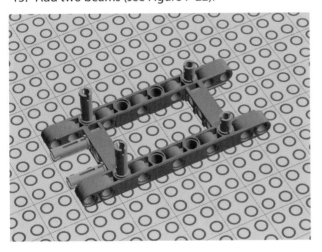

Figure 7-21. *Step 14: Preparing the frame*

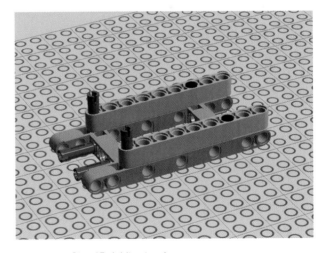

Figure 7-22. *Step 15: Adding two beams*

16. Add two connector pegs to the beams. Figure 7-23 shows how they go in.

17. Connect four 3M beams with pegs (see Figure 7-24).

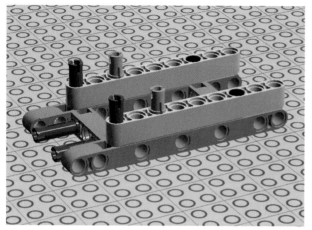

Figure 7-23. *Step 16: Connecting some pegs*

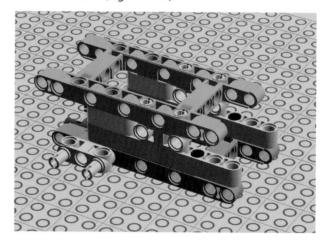

Figure 7-24. *Step 17: Connecting the 3M beams with pegs*

18. Then, throw on three 3M beams as shown in Figure 7-25.

19. Add another 5x11 Technic frame (Figure 7-26).

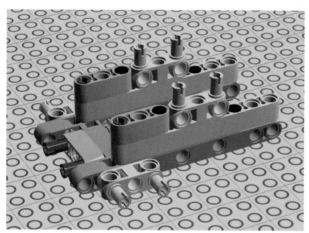

Figure 7-25. *Step 18: Adding more beams*

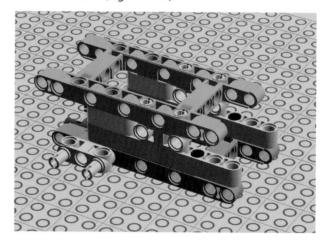

Figure 7-26. *Step 19: Attaching another frame*

20. Insert a pair of pegs as shown in Figure 7-27.

21. Add two 5M beams (Figure 7-28).

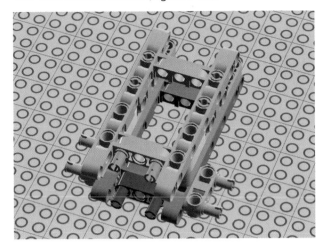

Figure 7-27. *Step 20: Attaching some pegs*

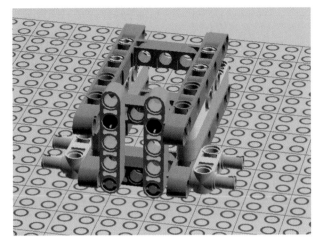

Figure 7-28. *Step 21: Adding two beams*

22. Grab a motor and add a 3M beam with pegs to it. Figure 7-29 shows this.

23. Connect a 5x7 Technic frame and a 3M beam with pegs as shown in Figure 7-30.

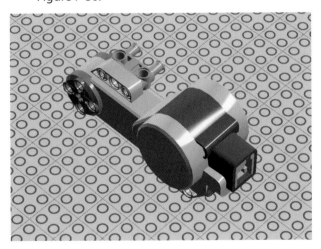

Figure 7-29. *Step 22: Connecting the motor*

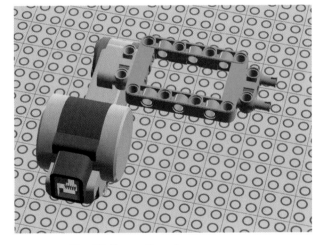

Figure 7-30. *Step 23: Connecting a frame*

24. Then, let's throw it onto the assembly you built in Steps 11 through 18 (see Figure 7-31).

25. Add another motor (Figure 7-32)!

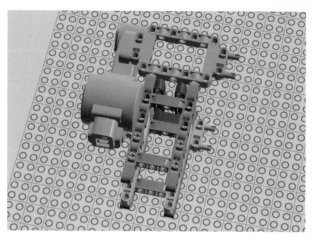

Figure 7-31. *Step 24: Attaching it to your earlier work*

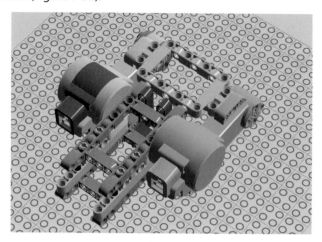

Figure 7-32. *Step 25: Adding another motor*

26. Add three pegs to the motor, as shown in Figure 7-33.

27. Then, add three more pegs to the other motor (see Figure 7-34).

Figure 7-33. *Step 26: Adding pegs to the motor*

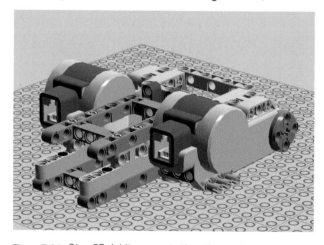

Figure 7-34. *Step 27: Adding pegs to the other motor*

28. Add a pair of beams to the pegs you just inserted. Figure 7-35 shows this.

29. Add four pegs as you see in Figure 7-36.

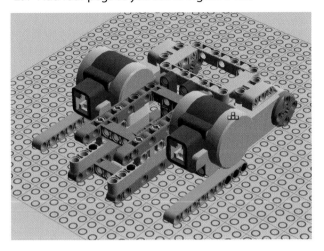

Figure 7-35. *Step 28: Adding two beams to the assembly*

Figure 7-36. *Step 29: Adding yet more pegs*

30. Then, four more pegs to the opposite side! Figure 7-37 shows this.

31. Connect two T-beams (see Figure 7-38).

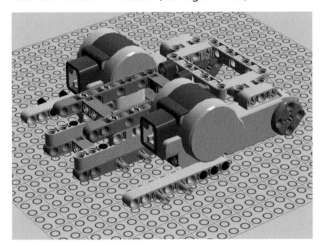

Figure 7-37. *Step 30: And some pegs for the other side*

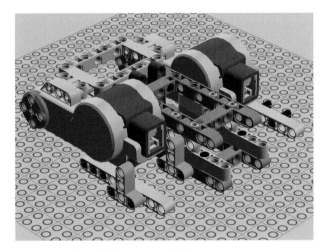

Figure 7-38. *Step 31: Connecting the T-beams*

32. Add two more T-beams to the opposite side, as shown in Figure 7-39.

33. Add two cross axles and secure them with bushes. Figure 7-40 shows this.

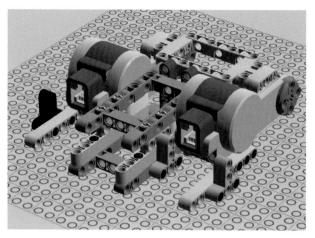

Figure 7-39. *Step 32: Adding the other T-beams*

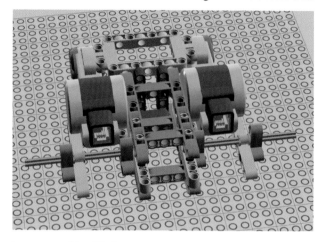

Figure 7-40. *Step 33: Connecting and securing the axles*

34. Insert cross axles into the motors' hubs and add two bushes to each axle (see Figure 7-41).

35. Throw on some rims to make it look like a car (see Figure 7-42)!

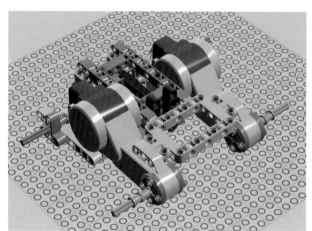

Figure 7-41. *Step 34: Adding axles to the motors' hubs*

Figure 7-42. *Step 35: Attaching the rims*

36. Add bushes to the rims' axles. Do the same to the other side. Figure 7-43 shows this.

37. Add four cross connectors as shown in Figure 7-44.

Figure 7-43. *Step 36: Adding the bushes*

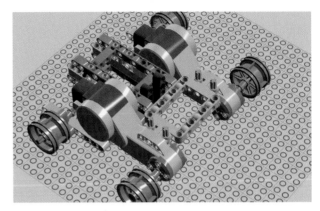

Figure 7-44. *Step 37: Adding four cross connectors*

38. Grab four angle elements and connect them as you see in Figure 7-45 with 3M cross axles and bushes. They won't want to stick together, so you may have to hold them in place.

39. Connect cross connectors, eight in all (see Figure 7-46).

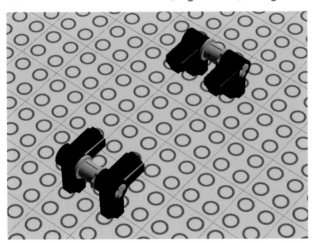

Figure 7-45. *Step 38: Preparing the angle elements*

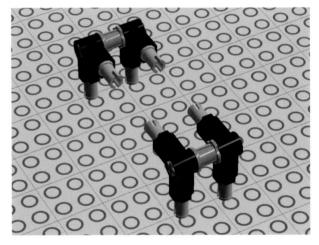

Figure 7-46. *Step 39: Adding cross connectors*

40. Connect the assemblies you just built to a turntable, as shown in Figure 7-47

41. Now, connect the turntable assembly to the car as shown in Figure 7-48.

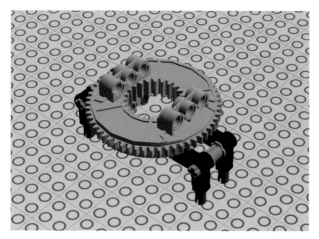

Figure 7-47. *Step 40: Attaching the angle assembly to the turntable*

Figure 7-48. *Step 41: Connecting the turntable to the car*

42. Add six connector pegs to the turntable (Figure 7-49).

43. Add two 5x7 Technic frames to the turntable. See Figure 7-50 for how this goes together.

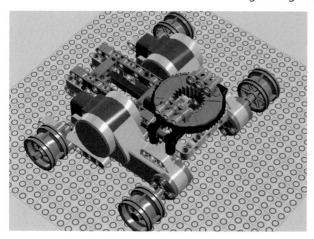

Figure 7-49. *Step 42: Adding some pegs*

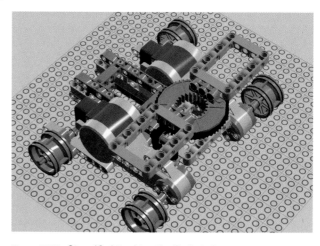

Figure 7-50. *Step 43: Attaching the Technic frames*

Chapter 7

44. Connect a 5M beam to one of the Technic frames with two 3M connector pegs (Figure 7-51).

45. Pop in two cross connectors as shown in Figure 7-52.

Figure 7-51. *Step 44: Connecting a beam with pegs*

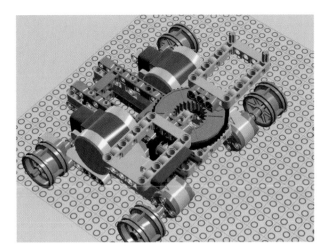

Figure 7-52. *Step 45: Adding two cross connectors*

46. Then add connector pegs, ten in all. Figure 7-53 shows the assembly so far.

47. The 15M beams added in Figure 7-54 will help stabilize the car's rotating platform.

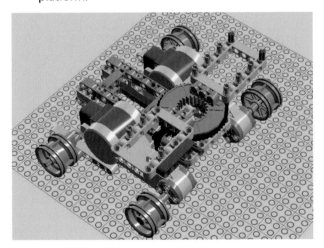

Figure 7-53. *Step 46: Adding connector pegs*

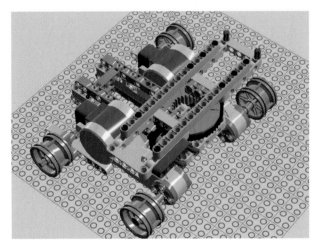

Figure 7-54. *Step 47: Adding two beams for stability*

48. Insert two cross connectors as shown in Figure 7-55.

49. Add a double cross block (see Figure 7-56).

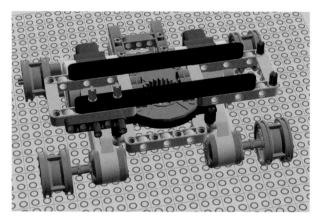

Figure 7-55. *Step 48: Adding cross connectors*

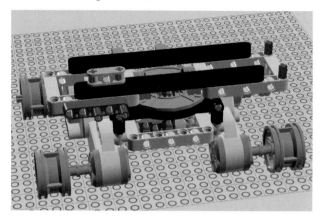

Figure 7-56. *Step 49: Attaching a double cross block*

50. Rest a motor on top of the cross block. The holes to the motor's left will line up with the double cross block as shown in Figure 7-57.

51. Add two connector pegs (Figure 7-58).

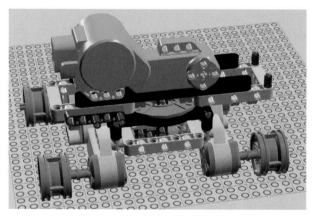

Figure 7-57. *Step 50: Placing the motor*

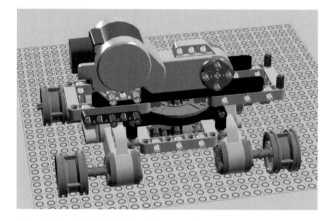

Figure 7-58. *Step 51: Adding the connector pegs*

52. Slide a Technic axle through the motor and double cross block, then secure one end with a bush. Figure 7-59 shows how it goes together.

53. Secure the other end of the cross axle with a bush (Figure 7-60).

Figure 7-59. *Step 52: Adding and securing the axle*

Figure 7-60. *Step 53: Securing the cross axle*

54. Attach a pair of 3M beams as shown in Figure 7-61.

55. Add a connector peg and cross connector to the motor's hub (see Figure 7-62).

Figure 7-61. *Step 54: Adding two beams*

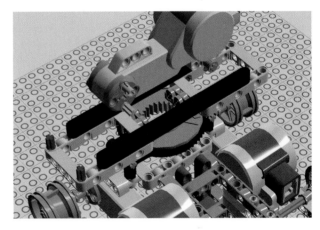

Figure 7-62. *Step 55: Adding peg and axle*

56. Attach a double angle beam to the pegs you just added (Figure 7-63).

57. Add a cross axle to the double angle beam, then add four bushes, as shown in Figure 7-64.

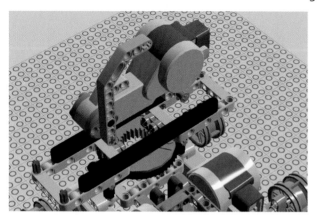

Figure 7-63. *Step 56: Adding a double angle beam*

Figure 7-64. *Step 57: Adding a cross axle and bushes*

58. You're done with this part! It should now look like Figure 7-65. Next, we'll work on the Gripperbot's arm.

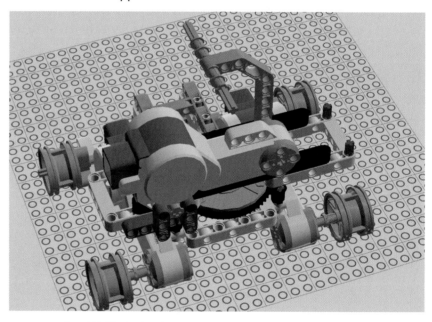

Figure 7-65. *Step 58: The body is finished...for now!*

59. Let's begin with one of the arm's claws. Add two axles to a double angle beam, then secure the bottoms with bushes as shown in Figure 7-66.

60. Add a 3M Techic lever (Figure 7-67).

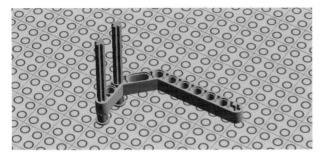

Figure 7-66. *Step 59: Attaching and securing axles to a beam*

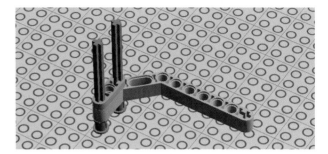

Figure 7-67. *Step 60: Adding the lever*

61. Then, throw on three 3M beams (see Figure 7-68).

62. Add a cross axle to the other end of the beam, and secure it with a half bush (Figure 7-69).

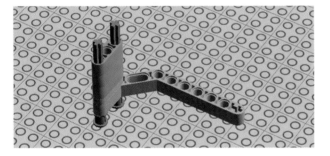

Figure 7-68. *Step 61: Attaching three beams*

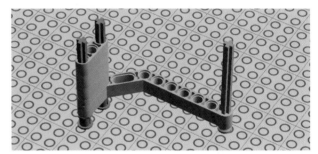

Figure 7-69. *Step 62: Adding another cross-axle*

63. Throw another half bush on there for good measure, as shown in Figure 7-70.

64. Add a double cross block (see Figure 7-71).

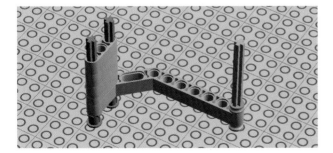

Figure 7-70. *Step 63: Securing it with a half bush*

Figure 7-71. *Step 64: Attaching a double cross block*

65. Secure the double cross block with a bush. Figure 7-72 shows this.

66. Add a 24-tooth gear as shown in Figure 7-73.

Figure 7-72. *Step 65: Securing the block*

Figure 7-73. *Step 66: Adding a gear*

67. Next, throw on another double angle beam (see Figure 7-74).

68. Secure the double angle beam with a bush and 3M lever (Figure 7-75).

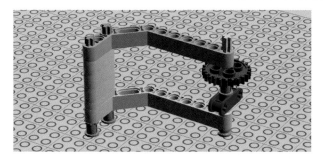

Figure 7-74. *Step 67: Adding another double angle beam*

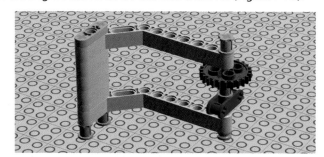

Figure 7-75. *Step 68: Putting the bush and lever in place*

69. Next, build another one just like it, but flip the double angle beams to make it a mirror image of the first one (as in Figure 7-76). You have claws!

70. Connect two Technic bricks with a single connector peg, as shown in Figure 7-77.

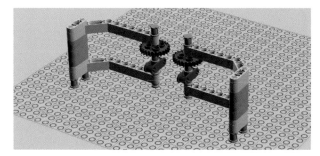

Figure 7-76. *Step 69: Building another*

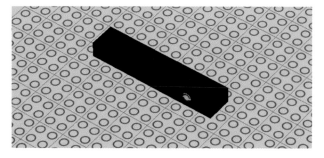

Figure 7-77. *Step 70: Connecting two Technic bricks*

Chapter 7

71. Add two 6M cross axles and secure them with a pair of double cross blocks on each side (Figure 7-78).

72. Add a 3M cross axle and secure it on one end with a 3x3 lever, as shown in Figure 7-79.

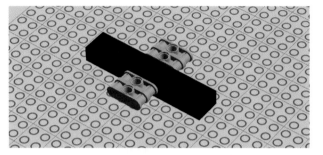

Figure 7-78. *Step 71: Adding cross axles*

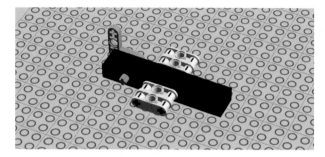

Figure 7-79. *Step 72: Adding another cross axle and securing it with a lever*

73. Add another 3x3 lever to the other side. Figure 7-80 shows the assembly so far.

74. Slide a 5M cross axle through and secure with bushes (Figure 7-81).

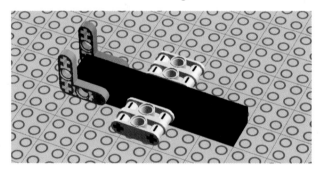

Figure 7-80. *Step 73: Adding another lever*

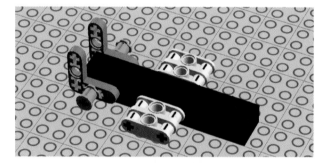

Figure 7-81. *Step 76: Inserting the cross axle*

75. Add a turntable and secure it with a pair of 2M cross axles (the red ones; they are shown as green in Figure 7-82 because green indicates a newly added part).

76. Add another cross axle. Figure 7-83 shows this.

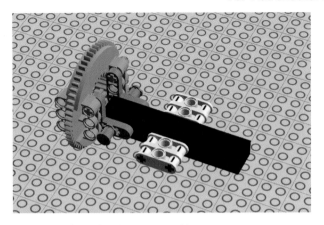

Figure 7-82. *Step 75: Adding a turntable*

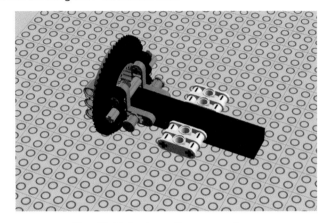

Figure 7-83. *Step 76: Sliding in another cross axle*

77. Add six connector pegs (Figure 7-84).

78. OK, you're done with this part! It should look like Figure 7-85. Set it aside for now.

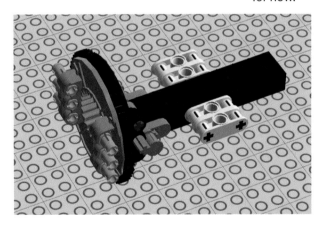

Figure 7-84. *Step 77: Adding connector pegs*

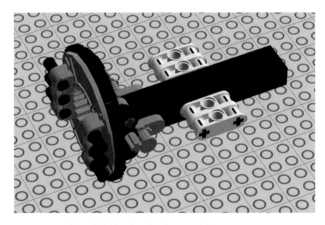

Figure 7-85. *Step 78: The finished assembly*

Chapter 7

79. Insert two 3M connector pegs and two regular connector pegs as shown in Figure 7-86.

80. Add three 5M beams (Figure 7-87).

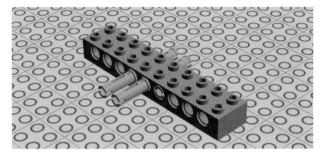

Figure 7-86. *Step 79: Adding pegs*

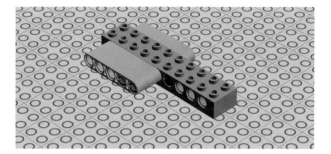

Figure 7-87. *Step 80: Adding beams*

81. Add two connector pegs to each side (see Figure 7-88).

82. Next, attach the assembly from Step 75 as shown in Figure 7-89.

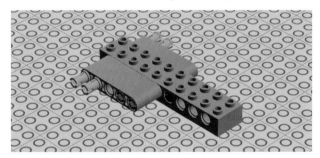

Figure 7-88. *Step 81: More pegs*

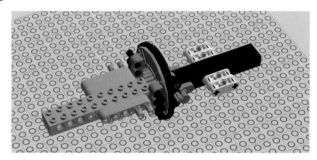

Figure 7-89. *Step 82: Attaching the assembly you built earlier*

83. Then, secure the two parts with a pair of T-beams (see Figure 7-90).

84. Grab a Power Functions motor, add a cross axle, then throw on a bush and an 8-tooth gear. Figure 7-91 shows how it looks now.

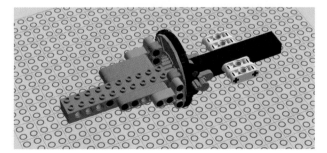

Figure 7-90. *Step 83: Securing it with T-beams*

Figure 7-91. *Step 84: Preparing the power functions motor*

85. Attach the motor to the arm assembly; the motor's 8-tooth gear should mesh with the turntable's teeth. Figure 7-92 shows the assembly.

86. Add a pair of 4M levers secured with a 3M cross axle as shown in Figure 7-93.

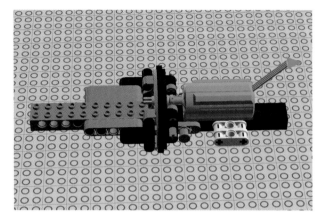

Figure 7-92. *Step 85: Attaching the motor to the arm*

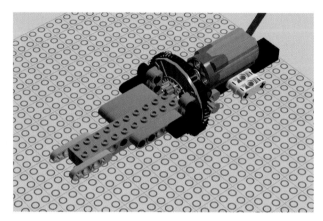

Figure 7-93. *Step 86: Adding lever beams*

87. Slide a 5M cross axle through; it won't want to stay there. Figure 7-94 shows how things should be arranged now.

88. Add a claw to each side as shown in Figure 7-95.

Figure 7-94. *Step 87: Adding a cross axle*

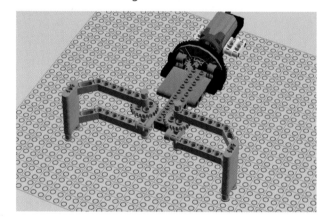

Figure 7-95. *Step 88: Adding the claws*

89. Secure a cross block with a pair of half bushes and a 5M cross axle (see Figure 7-96).

90. Add a cross axle and worm gear to the other Power Functions motor. Figure 7-97 shows this.

Figure 7-96. *Step 89: Securing a cross block*

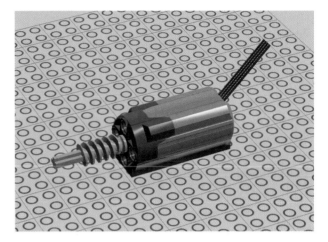

Figure 7-97. *Step 90: Connecting a worm gear to the other motor*

91. Slide the motor's axle through the cross block's upper hole, then secure the motor to the Technic bricks' studs as shown in Figure 7-98.

92. Slide a 7M cross axle through and secure it with a half bush on each side (see Figure 7-99).

Figure 7-98. *Step 91: Attaching the motor*

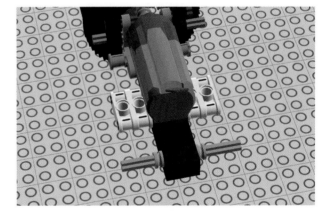

Figure 7-99. *Step 92: Adding a cross axle*

93. Add angle elements and bushes. Figure 7-100 shows this.

94. Attach the completed arm to the Gripperbot's rotating platform as shown in Figure 7-101.

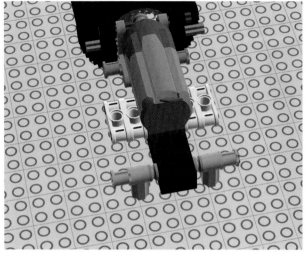

Figure 7-100. *Step 93: Adding angle elements*

Figure 7-101. *Step 94: Attaching the arm*

95. Add a 9M beam to the arm, using the holes in front of the quartet of cross blocks. Secure the axle with four bushes. Figure 7-102 shows this.

96. Connect two beams (see Figure 7-103).

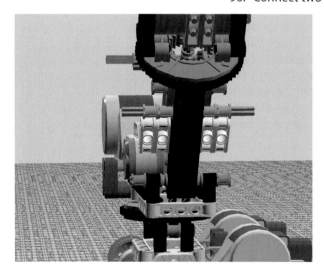

Figure 7-102. *Step 95: Adding a beam and securing the axle*

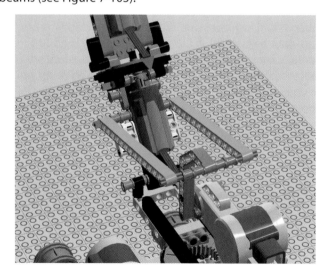

Figure 7-103. *Step 96: Connecting two beams*

97. Secure the beams with four half-bushes as shown in Figure 7-104.

98. Add a pair of connector pegs and a 3x5 90-degree angle beam (Figure 7-105).

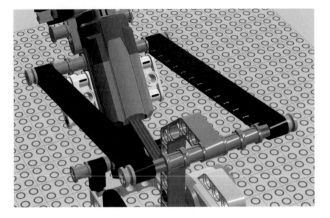

Figure 7-104. *Step 97: Securing the beams*

Figure 7-105. *Step 98: Adding an angle beam*

99. Add a pair of connector pegs as seen in Figure 7-106.

100. Add a pair of cross blocks, secured with two 6M axles as shown in Figure 7-107.

Figure 7-106. *Step 99: Two more pegs*

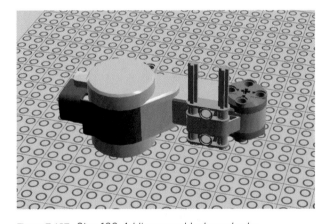

Figure 7-107. *Step 100: Adding cross blocks and axles*

101. Insert another 6M axle in the motor's hub (Figure 7-108).

102. Add a 16-tooth gear and a pair of bushes. Figure 7-109 shows this.

Figure 7-108. *Step 101: Inserting a cross axle*

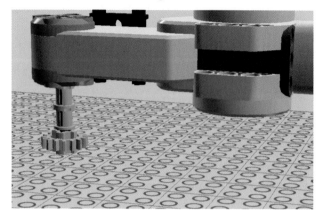

Figure 7-109. *Step 102: Adding a gear and two bushes*

103. Attach the motor assembly to the main robot, connecting to the two pegs you added in Step 98 (Figure 7-110).

104. Shove down the two cross axles you added in Step 100 and add a pair of half bushes. When you're done it should look Figure 7-111.

Figure 7-110. *Step 103: Attaching the motor assembly*

Figure 7-111. *Step 104: Shoving down the axles and securing with bushes*

Chapter 7

105. Add two pegs as shown in Figure 7-112.

106. Add an 11M beam and three cross connectors as seen in Figure 7-113. This beam will hold the Bricktronics Motor Controller.

Figure 7-112. *Step 105: Adding more pegs*

Figure 7-113. *Step 106: Add a beam and cross connectors*

107. Add two 3M beams with pegs. Figure 7-114 shows this.

108. Attach two pegs and two 2M beams with cross hole (Figure 7-115).

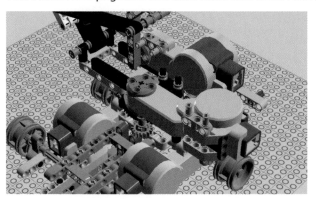

Figure 7-114. *Step 107: Adding a pair of 3M beams with pegs*

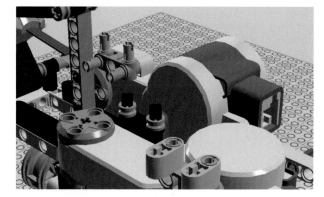

Figure 7-115. *Step 108: Adding pegs and 2M Technic beams*

109. Add a pair of tubes and two 3M cross axles (Figure 7-116).

110. Add a two cross connectors and two 3M Technic levers as shown in Figure 7-117.

Figure 7-116. *Step 109: Adding tubes and 3M cross axles*

Figure 7-117. *Step 110: Adding cross connectors and 3M Technic levers*

111. Secure the ends of a pair of 6M cross axles with a 2M Technic lever. Figure 7-118 shows this. These connectors will accommodate the battery pack for the turret.

112. Slide a pair of 3M cross blocks down the axles (Figure 7-119).

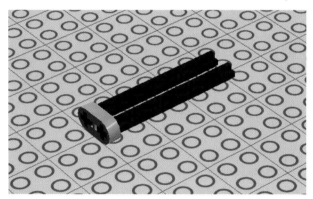

Figure 7-118. *Step 111: Securing the ends of a pair of axles*

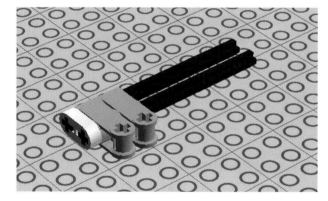

Figure 7-119. *Step 112: Sliding on a pair of cross blocks*

113. Add three 9M beams as shown in Figure 7-120.

114. Add two more cross blocks and then top it off with a 2M Technic lever (Figure 7-121).

Figure 7-120. *Step 113: Adding beams*

Figure 7-121. *Step 114: More cross blocks and a Technic lever*

115. Add the assembly you just completed and secure it with a pair of 8M axles as shown in Figure 7-122.

116. Add two more cross blocks to the 8M axles and then secure the ends with 2M levers (Figure 7-123).

Figure 7-122. *Step 115: Adding the assembly*

Figure 7-123. *Step 116: Securing the assembly*

117. Add six 2M cross axles. Your Arduino's plate will mount to these, as shown in Figure 7-124.

Figure 7-124. *Step 117: Adding 2M cross axles*

WHERE ARE THE TANK TREADS?

Lego Digital Designer, which we used to create the step-by-steps for this book, can't manage tank treads so we've omitted the treads from the instructions. Just throw them on! If they're a little tight, just give them a yank to stretch them out a bit.

Assembling the Gripperbot's Electronics

The system concept for the Gripperbot is relatively simple. There are a variety of motors. The base has two of them, the right and left tank treads. The turret has four of them, the spinner, the "elbow," the "wrist," and the gripper. The bracers translate motion and button presses into speeds for specific motors. All of the XBee nodes broadcast to all other XBee nodes. The base and turret receive all messages, and only act on the ones that are intended for motors that they control. The turret uses the Bricktronics Motor Controller, which comes pre-programmed with a set of commands for motors 1 through 5, one for each of its ports. This project only uses four motors on the Motor Controller, which are numbered motors 1 through 4, so the two tread motors on the base's Arduino are numbered 6 and 7.

The base of the Gripperbot packs an Arduino, an XBee shield with XBee, and a Bricktronics Shield, as seen in Figure 7-125. The left and right motors plug into the Bricktronics Shield's motor ports. A 6-AA battery pack powers this portion of the robot.

Figure 7-125. *The base of the robot is controlled by a Bricktronics Shield*

The arm portion of the robot (Figure 7-126) has a Bricktronics Motor Board equipped with an XBee. The Power Functions motors plug into the Motor Board via Molex connectors (see "Adding Molex Connectors to Lego Wires" in Chapter Ten) and the two Mindstorms motors plug directly into the board via Mindstorms wires. Another 6-AA battery pack powers the arm.

If your battery packs didn't come with DC plugs, attach one like the one in the parts list. Remember to solder the positive wire of the battery pack to the central terminal, and the negative wire to the outer terminal.

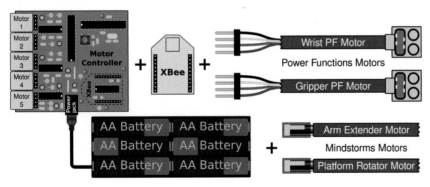

Figure 7-126. *The Arm*

Programming the Gripperbot

Here's the sketch for the Gripperbot base:

```
#include <Wire.h>
#include <Adafruit_MCP23017.h>
#include <Bricktronics.h>

// Make: Lego and Arduino Projects
// Chapter 7: Gripperbot: Base
// Website: http://www.wayneandlayne.com/bricktronics

Bricktronics brick = Bricktronics();
Motor r = Motor(&brick, 1);
Motor l = Motor(&brick, 2);

void setup() {❶
  Serial.begin(9600);
  brick.begin();
  r.begin();
  l.begin();
}

void process_incoming_command(char cmd, char arg0) ❷
{
  int speed = 0;
  switch (cmd)
  {
  case '6':
  case 6: ❸
    speed = arg0*2;
    l.set_speed(speed);
    break;
  case '7':
  case 7: ❹
    speed = arg0*2;
    r.set_speed(speed);
    break;
  default:
    break;
  }
}

void loop() {
  while (Serial.available() < 3)
  {
    //Do nothing while we wait for a full command to come in.
  }
  char start = Serial.read(); ❺
  if (start != '*') ❻
  {
    return; ❼
  }

  char cmd = Serial.read(); ❽
  char arg0 = Serial.read();
  process_incoming_command(cmd, arg0);
}
```

❶ The `setup()` function runs once at startup.

❷ `process_incoming_command()` takes in the two bytes of the body of a command, and then performs the specified action.

❸ If the command was "set the speed to motor 6 (left tank tread)," then the code that follows is run.

❹ If the command was "set the speed to motor 7 (right tank tread)," the code that follows is run.

❺ Remove the oldest character from the serial buffer, and put it into start.

❻ This isn't the start of the command, so let's try the next one.

❼ Start back over at the top of `loop()`.

❽ Read the command and argument, and call `process_incoming_command()`.

Here's the sketch for the Gripperbot left bracer:

```
#include <Wire.h>
#include <ArduinoNunchuk.h>  ❶

// Make: Lego and Arduino Projects
// Chapter 7: Gripperbot: Left Bracer  ❷
// Website: http://www.wayneandlayne.com/bricktronics/


ArduinoNunchuk nunchuk = ArduinoNunchuk();

#define NUM_OF_READINGS 10  ❸

char pitch_index = 0;
int pitch_readings[NUM_OF_READINGS];
int pitch_total = 0;

void setup()  ❹
{
  Serial.begin(9600);
  init_nunchuk_connections();
  delay(100);
  nunchuk.init();

  for (int i = 0; i < NUM_OF_READINGS; i++)  ❺
  {
    pitch_readings[i] = 0;
  }
}

void loop()  ❻
{
  nunchuk.update();  ❼
  handle_left_tank_tread(nunchuk.accelY);
  delay(50);
}
```

```
                              void init_nunchuk_connections()  ❽
                              {
                                pinMode(A2, OUTPUT);
                                digitalWrite(A2, LOW);
                                pinMode(A3, OUTPUT);
                                digitalWrite(A3, HIGH);
                              }

                              int average_accel_pitch(int raw_pitch)  ❾
                              {

                                pitch_total -= pitch_readings[pitch_index];  ❿
                                pitch_readings[pitch_index] = raw_pitch;  ⓫
                                pitch_total += pitch_readings[pitch_index];  ⓬
                                pitch_index += 1;  ⓭

                                if (pitch_index >= NUM_OF_READINGS)  ⓮
                                {
                                  pitch_index = 0;
                                }

                                return pitch_total / NUM_OF_READINGS;  ⓯
                              }

                              int last_left_speed = 0;
                              void handle_left_tank_tread(int raw_accel_pitch)  ⓰
                              {
                                int avg_pitch = average_accel_pitch(raw_accel_pitch);
                                #define PITCH_CENTER 550  ⓱
                                #define MAX_PITCH PITCH_CENTER+150
                                #define MIN_PITCH PITCH_CENTER-150

                                #define MAX_FORWARD 127  ⓲
                                #define MAX_BACKWARD -128
                                avg_pitch = constrain(avg_pitch, MIN_PITCH, MAX_PITCH);
                                int left_speed = map(avg_pitch, MIN_PITCH, MAX_PITCH,
                                                      MAX_BACKWARD, MAX_FORWARD);  ⓳
                                left_speed = constrain(left_speed, MAX_BACKWARD, MAX_FORWARD);  ⓴

                                if (left_speed != last_left_speed)  ㉑
                                {
                                  last_left_speed = left_speed;
                                  Serial.write('*');  ㉒
                                  Serial.write(6);  ㉓
                                  char out = left_speed;  ㉔
                                  Serial.write(out);  ㉕
                                }
                              }
```

❶ While there are plenty of Nunchuk libraries, the one we used can be downloaded at *http://www.gabrielbianconi.com/projects/arduinonunchuk/*.

❷ In the default configuration, the left bracer controls the left tank tread. When the Nunchuk is tilted forward, the left tank tread goes forward. When the Nunchuk is tilted back, the left tank tread goes backward.

❸ NUM_OF_READINGS define controls how many samples are taken in the running average.

❹ `setup()` runs once at startup.

❺ Initialize the variables for the pitch averaging.

❻ `loop()` runs repeatedly until power is removed.

❼ Updating the nunchuk object reads new data from the sensors.

❽ The Solarbotics NunChucky plugs the Nunchuk's 4 connections into the lower analog pins. The Nunchuk only needs a little power, so we can actually power it from the Arduino pins!

❾ `average_accel_pitch` makes a running average of the last `NUM_OF_READINGS` readings from the Nunchuk. This is useful because the accelerometer on the Nunchuk can be a little noisy, and a very simple way of removing some of the noise is to do a running average. See the Arduino example sketch "Smoothing" for more details on running averages.

❿ Remove the oldest pitch reading from the total.

⓫ Replace the oldest pitch reading with the current pitch reading.

⓬ Add the newest pitch reading to the total.

⓭ Advance the index to the next spot.

⓮ If the index is past the end, bring it back to zero.

⓯ Return the average by dividing the total by the number of readings.

⓰ `handle_left_tank_tread()` takes in the Nunchuk's pitch value, and sends a command, if needed, to the Arduino on the body telling it the speed to move the left tank tread.

⓱ The readings out of the Nunchuk are centered at approximately 550. It goes to about 700 forward, and to about 300 backward. These may vary on your unit a little—if so, feel free to adjust the constants to make it more accurate on your Nunchuk. We're going to translate this range to a signed range with 0 pitch being straight up, forward being positive, and back being negative.

⓲ We've picked this range because it is the maximum range that can fit in a signed byte.

⓳ This maps our number from the range `MIN_PITCH` to `MAX_PITCH` to the range `MAX_BACKWARD` to `MAX_FORWARD`.

⓴ No matter what, our number is going to range from `MAX_BACKWARD` to `MAX_FORWARD`.

㉑ There are a lot of things you can do here, to keep the system responsive but cut down on unnecessary radio traffic, which takes power. You can only update if the change goes over a threshold, or add more fancy filtering, but we're going to go with a simple option of "transmit the speed if it isn't the same as the last speed we transmitted."

㉒ In our system, all commands start with an asterisk.

㉓ The second byte of the command is the motor number. The left motor is 6. `Serial.write()` transmits the byte without making it "printable."

㉔ Even though `left_speed` currently is between 127 and -128, which fits in a char, we need to make it a char so we only send one byte.

㉕ The third byte of the command is the speed, from -128 to 127.

Here's the last sketch, for the Gripperbot right bracer:

```
#include <Wire.h>
#include <ArduinoNunchuk.h>  ❶

// Make: Lego and Arduino Projects
// Chapter 7: Gripperbot: Right Hand Bracer
// Website: http://www.wayneandlayne.com/bricktronics

ArduinoNunchuk nunchuk = ArduinoNunchuk();

#define NUM_OF_READINGS 10  ❷

char roll_index = 0;
int roll_readings[NUM_OF_READINGS];
int roll_total = 0;
char pitch_index = 0;
int pitch_readings[NUM_OF_READINGS];
int pitch_total = 0;

void setup()  ❸
{
  Serial.begin(9600);
  init_nunchuk_connections();
  delay(100);
  nunchuk.init();

  for (int i = 0; i < NUM_OF_READINGS; i++)  ❹
  {
    roll_readings[i] = 0;
    pitch_readings[i] = 0;
  }
}

void loop()  ❺
{
  nunchuk.update();  ❻
  handle_right_tank_tread(nunchuk.accelY);
  handle_wrist(nunchuk.accelX);
  handle_spin(nunchuk.analogX);
  handle_elbow(nunchuk.analogY);
  handle_grip(nunchuk.zButton, nunchuk.cButton);

  delay(50);
}

void init_nunchuk_connections()  ❼
{
  pinMode(A2, OUTPUT);
  digitalWrite(A2, LOW);
  pinMode(A3, OUTPUT);
  digitalWrite(A3, HIGH);
}
```

```
int average_accel_pitch(int raw_pitch) ❽
{
  pitch_total -= pitch_readings[pitch_index]; ❾
  pitch_readings[pitch_index] = raw_pitch; ❿
  pitch_total += pitch_readings[pitch_index]; ⓫

  pitch_index += 1; ⓬

  if (pitch_index >= NUM_OF_READINGS) ⓭
  {
    pitch_index = 0;
  }

  return pitch_total / NUM_OF_READINGS; ⓮
}

int average_accel_roll(int raw_roll) ⓯
{
  roll_total -= roll_readings[roll_index];
  roll_readings[roll_index] = raw_roll;
  roll_total += roll_readings[roll_index];

  roll_index += 1;
  if (roll_index >= NUM_OF_READINGS)
  {
    roll_index = 0;
  }

  return roll_total / NUM_OF_READINGS;
}

int last_right_speed = 0;
void handle_right_tank_tread(int raw_accel_pitch) ⓰
{
  int avg_pitch = average_accel_pitch(raw_accel_pitch);


  #define PITCH_CENTER 550 ⓱
  #define MAX_PITCH PITCH_CENTER+150
  #define MIN_PITCH PITCH_CENTER-150

  #define MAX_FORWARD 127 ⓲
  #define MAX_BACKWARD -128

  avg_pitch = constrain(avg_pitch, MIN_PITCH, MAX_PITCH);
  int right_speed = map(avg_pitch, MIN_PITCH,
                        MAX_PITCH, MAX_BACKWARD, MAX_FORWARD); ⓳
  right_speed = constrain(right_speed, MAX_BACKWARD, MAX_FORWARD); ⓴

  if (right_speed != last_right_speed) ㉑
  {
    last_right_speed = right_speed;
    Serial.write('*'); ㉒
    Serial.write(7); ㉓
    char out = right_speed; ㉔
    Serial.write(out); ㉕
  }
}
```

```
int last_spin_speed = 0;
void handle_spin(int joy_x) ㉖
{
  #define JOY_X_CENTER 129 ㉗

  #define MAX_JOY_X JOY_X_CENTER+101
  #define MIN_JOY_X JOY_X_CENTER-97

  #define MAX_RIGHT_SPIN 127
  #define MAX_LEFT_SPIN -128

  joy_x = constrain(joy_x, MIN_JOY_X, MAX_JOY_X);
  int spin_speed = map(joy_x, MIN_JOY_X, MAX_JOY_X,
                       MAX_LEFT_SPIN, MAX_RIGHT_SPIN);
  spin_speed = constrain(spin_speed, MAX_LEFT_SPIN, MAX_RIGHT_SPIN);

  if (spin_speed != last_spin_speed)
  {
    last_spin_speed = spin_speed;
    Serial.write('*');
    Serial.write(1);
    char out = spin_speed;
    Serial.write(out);
  }
}

int last_elbow_speed = 0;
void handle_elbow(int joy_y) ㉘
{
  #define JOY_Y_CENTER 126 ㉙

  #define MAX_JOY_Y JOY_Y_CENTER+86
  #define MIN_JOY_Y JOY_Y_CENTER-98

  #define MAX_UP_ELBOW 127
  #define MAX_DOWN_ELBOW -128

  joy_y = constrain(joy_y, MIN_JOY_Y, MAX_JOY_Y);
  int elbow_speed = map(joy_y, MIN_JOY_Y, MAX_JOY_Y,
                        MAX_UP_ELBOW, MAX_DOWN_ELBOW);
  elbow_speed = constrain(elbow_speed, MAX_DOWN_ELBOW, MAX_UP_ELBOW);

  if (elbow_speed != last_elbow_speed)
  {
    last_elbow_speed = elbow_speed;
    Serial.write('*');
    Serial.write(2);
    char out = elbow_speed;
    Serial.write(out);
  }
}

int last_wrist_speed = 0;
void handle_wrist(int raw_accel_roll) ㉚
{
  int avg_roll = average_accel_roll(raw_accel_roll);
  #define ROLL_CENTER 560 ㉛

  #define MAX_ROLL ROLL_CENTER+220
  #define MIN_ROLL ROLL_CENTER-220
```

```
#define MAX_RIGHT_WRIST 127
#define MAX_LEFT_WRIST -128

avg_roll = constrain(avg_roll, MIN_ROLL, MAX_ROLL);
int wrist_speed = map(avg_roll, MIN_ROLL, MAX_ROLL,
                      MAX_LEFT_WRIST, MAX_RIGHT_WRIST);
wrist_speed = constrain(wrist_speed, MAX_LEFT_WRIST, MAX_RIGHT_WRIST);

if (wrist_speed != last_wrist_speed)
{
  last_wrist_speed = wrist_speed;
  Serial.write('*');
  Serial.write(3);
  char out = wrist_speed;
  Serial.write(out);
}
}

char last_cmd = 0;
void handle_grip(boolean z, boolean c) ❷
{
  #define GRIP_OUT_SPEED 100
  #define GRIP_IN_SPEED -100

  if (z == c)
  {
    if (last_cmd != 0)
    {
      last_cmd = 0;
      Serial.write('*');
      Serial.write(4);
      Serial.write((unsigned byte) 0);
    }
  } else if (z) //out
  {
    if (last_cmd != 'z')
    {
      last_cmd = 'z';
      Serial.write('*');
      Serial.write(4);
      Serial.write(GRIP_OUT_SPEED);
    }
  } else if (c) //in
  {
    if (last_cmd != 'c')
    {
      last_cmd = 'c';
      Serial.write('*');
      Serial.write(4);
      Serial.write(GRIP_IN_SPEED);
    }
  }
}
```

❶ While there are plenty of Nunchuk libraries, the one we used can be downloaded at *http://www.gabrielbianconi.com/projects/arduinonunchuk/*

❷ NUM_OF_READINGS controls how many samples are taken in the running average.

❸ setup() runs once at startup.

❹ Initialize the variables for the pitch and roll averaging.

❺ loop() runs repeatedly until power is removed.

❻ Read new data from the Nunchuk.

❼ The Solarbotics NunChucky plugs the Nunchuk's four connections into the lower analog pins. The Nunchuk only needs a little power, so we can actually power it from the Arduino pins!

❽ average_accel_pitch() makes a running average of the last NUM_OF_READINGS readings from the Nunchuk. This is useful because the accelerometer on the Nunchuk can be a little noisy, and a very simple way of removing some of the noise is to do a running average. See the Arduino example sketch "Smoothing" (File | Examples | 03.Analog | Smoothing) for more details on running averages. This is also the same function as in the left bracer, and nearly the same as the averaging function that handles roll.

❾ Remove the oldest pitch reading from the total.

❿ Replace the oldest pitch reading with the current pitch reading.

⓫ Add the newest pitch reading to the total.

⓬ Advance the index to the next spot.

⓭ If the index is past the end, bring it back to zero.

⓮ Return the average by dividing the total by the number of readings.

⓯ average_accel_roll() averages the raw roll values just like average_accel_pitch() does.

⓰ handle_right_tank_tread() takes in the Nunchuk's pitch value, and sends a command, if needed, to the Arduino on the tank body telling it the speed to move motor 7, or the right tank tread.

⓱ The readings out of the Nunchuk are centered at approximately 550. They go up to about 700 tilted forward, and about 300 when tilted back. These may vary on your unit a little—if so, feel free to adjust the constants to make it more accurate on your Nunchuk. We're going to translate this range to a signed range with 0 pitch being straight up, forward being positive, and back being negative.

⓲ We've picked this range as the maximum range that can fit in a signed char, which is what we transmit.

⓳ This maps our number from the range MIN_PITCH to MAX_PITCH to the range MAX_BACKWARD to MAX_FORWARD.

⓴ No matter what, our number is going to range from MAX_BACKWARD to MAX_FORWARD.

Chapter 7

㉑ There are a lot of things you can do here, to keep the system responsive but cut down on unnecessary radio traffic, which takes power. You can only update if the change goes over a threshold, or add more fancy filtering, but we're going to go with a simple option of "transmit the speed if it isn't the same as the last speed we transmitted."

㉒ In our system, all commands start with an asterisk.

㉓ The second byte of the command is the motor number. The right tank tread motor is 7. `Serial.write()` transmits the byte without making it printable.

㉔ Even though `right_speed` currently is between 127 and -128, which fits in a `char`, we need to send a `char`, not an `int`, which is two bytes.

㉕ The third byte of the command is the speed, from -128 to 127.

㉖ `handle_spin()` takes in the Nunchuk's joystick's x value, and sends a command, if needed, to the motor controller telling it the speed to move motor 1, or the "spin" motor.

㉗ The joystick's x range is centered at 129, and it goes about 101 to the right, and about 97 to the left. These may vary on your unit a little—if so, feel free to adjust the constants to make it more accurate on your Nunchuk. We're going to translate this range to a signed range with 0 being straight up, right being positive, and left being negative.

㉘ `handle_elbow()` takes in the Nunchuk's joystick's y value, and sends a command, if needed, to the motor controller telling it the speed to move motor 2, or the "elbow."

㉙ The joystick's y range is centered at 126, and it goes about 86 forward, and about 98 backward. These may vary on your unit a little—if so, feel free to adjust the constants to make it more accurate on your Nunchuk. We're going to translate this range to a signed range with 0 being straight up, forward being positive, and backward being negative.

㉚ `handle_wrist()` takes in the Nunchuk's accelerometer's roll value, and sends a command, if needed, to the motor controller telling it the speed to move the motor 3, or the "wrist" motor.

㉛ The joystick's roll range is centered at 560, and it goes about 220 right, and 220 left. These may vary on your unit a little—if so, feel free to adjust the constants to make it more accurate on your Nunchuk. We're going to translate this range to a signed range with 0 being straight up, right being positive, and left being negative.

㉜ `handle_grip()` takes in the Nunchuk's accelerometer's roll value, and sends a command, if needed, to the motor controller telling it the speed to move motor 4, or the gripper motor.

The Next Chapter

In Chapter Eight we'll do another project: a Lego Keytar packing delicious electronic noisemakers! Are you starting a nerd band? This Keytar will be just the ticket.

Project: Keytar

8

Figure 8-1. *Got Lego and a Bricktronics Shield? Start a band, man!*

For our next project we're going to explore the realm of electronic music by building a working guitar-shaped Lego synthesizer! We call it a Keytar, because you mainly play it by pressing buttons. The name may be a stretch, but the project is fun to play with nonetheless!

This project creates crazy electronic music similar to the sounds generated by an Atari Punk Console, a classic electronic instrument of the earliest days of DIY electronic music (see the Atari Punk Console sidebar). Let's get building!

Parts List

This project has a large number of Lego parts as well as buttons, wires, and other electronic components. We'll also use new Bricktronics mounting plates (Figure 8-2) to accommodate the buttons, allowing them to slickly connect to the Keytar body.

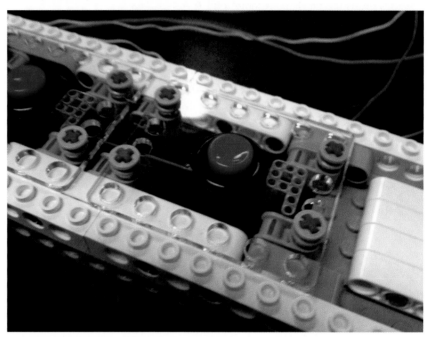

Figure 8-2. *We'll attach non-Lego buttons to the Keytar with the help of these nifty Brick-tronics mounting plates*

Tools & Electronics

You'll need the following electronics and associated parts to make the Keytar:

- Arduino Uno

- Bricktronics Shield

- Mounting plates for buttons and Arduino

- Power supply rated for 9V at 1.3A or greater with a 2.1mm center-positive plug. This provides power to the Arduino.

- 22-gauge solid-core wires (we used Maker Shed P/N MKEE3)

- 5 Momentary buttons (we used Jameco P/N 106112, but any SPST momentary on button will work)

- 1 piezo buzzer (we used Digi-key P/N PKM13EPYH4000-A0)

ATARI PUNK CONSOLE

Prolific electronics writer Forrest M. Mims III created a "Stepped Tone Generator" in 1980. It has since been "rebranded" as the "Atari Punk Console" (APC), but regardless of the name, it makes awesome noise. The APC works by combining the signals of two 555 timer ICs (integrated circuits) as adjusted by potentiometers.

The 555 was created in 1972 and is one of the widest-used ICs in existence with over a billion chips manufactured annually as late as 2003. Among its other uses, a 555 generates a repetitive, oscillating electronic signal that the APC uses to generate noises. The APC's potentiometers adjust the frequency of the oscillator and the width of the pulse, and this changes the sound generated by the 555s. In its most

basic form the APC's output can barely be called music—it's the sort of squawks and buzzes that drive parents up the wall and scare the cat away.

Of course, electronic hobbyists and musicians have created a vast variety of projects that have expanded on that original noisemaker, inspired to create amazing new forms of music out of electronic circuitry. For instance, circuit-bending website *http://GetLoFi.com* sells an APC kit (Figure 8-3) through their online store, as well as several other noisemakers that arguably draw inspiration from the original.

We're no different! The Keytar project doesn't actually use a 555 in it, but we have to tip our hats to the APC for inspiration.

Figure 8-3. *The Atari Punk Console is the classic electronic musical circuit, and it inspired the Keytar project. Credit: GetLoFi*

Lego Elements

So far in this book we've included mostly Mindstorms parts in the robots. However, the Keytar uses a lot of Lego plates (Figure 8-4), used to give the instrument its body. Reference the following list when assembling the parts:

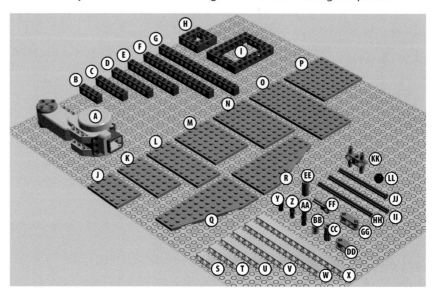

Figure 8-4. *Collect these parts to build the Keytar*

A. 2 Mindstorms motors

B. 3 Technic bricks 4M

C. 3 Technic bricks 6M

D. 2 Technic bricks 8M

E. 6 Technic bricks 10M

F. 16 Technic bricks 14M

G. 10 Technic bricks 16M

H. 4 Technic bricks 4x4

I. 2 Technic bricks 6x8

J. 1 Lego plate 4x6

K. 5 Lego plates 4x8

L. 2 Lego plates 4x10

M. 2 Lego plates 6x8

N. 3 Lego plates 6x10

O. 5 Lego plates 8x8

P. 9 Lego plates 6x14

Q. 1 Left wing 6x12

R. 1 Right wing 6x12

S. 2 Technic beams 3M

T. 4 Technic beams 5M

U. 1 Technic beam 7M

V. 15 Technic beams 9M

W. 1 Technic beam 13M

X. 1 Technic beam 15M

Y. 104 connector pegs

Z. 43 connector pegs w/cross axle

AA. 10 connector pegs 3M

BB. 8 bushes

CC. 2 cross axle extensions

DD. 20 cross blocks

EE. 6 tubes (optionally, use multiple colors)

FF. 2 pole reverser handles

GG. 3 double cross blocks

HH. 4 cross axles 9M

II. 2 cross axles 10M

JJ. 4 cross axles 8M w/end stop

KK. 14 3M beams w/pegs

LL. 2 ball elements

MM. 3 Mindstorms wires (not pictured)

Assembly Instructions

The Keytar (Figure 8-5) is a rather complicated build with numerous Lego parts needed, as well as electronic components like buttons and wires. Let's do it!

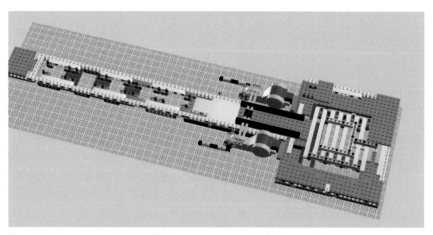

Figure 8-5. *Are you ready to build this awesome musical instrument?*

Build the Lego Model

The first step to creating the Keytar is to build the Lego chassis that supports the electronics, and looks groovy.

1. Let's begin by connecting two 15M beams with two 3M beams with pegs, as seen in Figure 8-6.

2. More 3M beams with pegs (Figure 8-7).

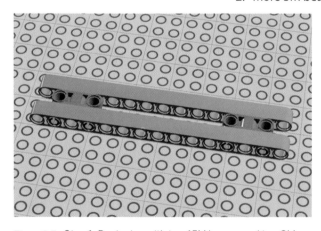

Figure 8-6. *Step 1: Beginning with two 15M beams and two 3M beams with pegs*

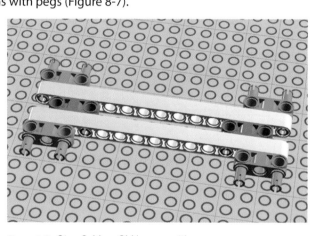

Figure 8-7. *Step 2: More 3M beams with pegs*

3. And more 15M beams (Figure 8-8)!

4. Those 3M beams with pegs are very useful! Let's add four more (Figure 8-9).

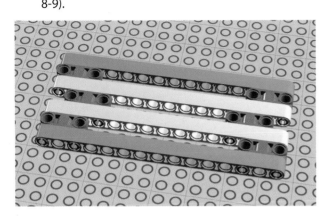

Figure 8-8. *Step 3: Adding two more 15M beams*

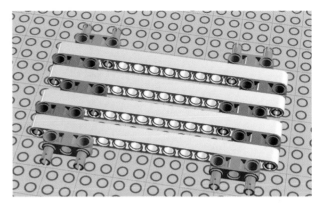

Figure 8-9. *Step 4: Adding more beams with pegs*

5. Connect two more 15M beams, as seen in Figure 8-10.

6. Then four more 3M beams with pegs (Figure 8-11)!

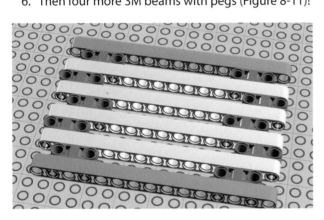

Figure 8-10. *Step 5: Two more 15M beams*

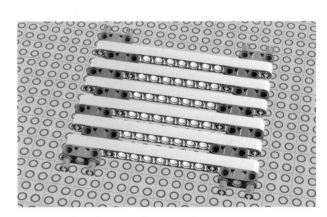

Figure 8-11. *Step 6: Four 3M beams with pegs*

7. Add two more 15M beams (Figure 8-12).

8. This step is kind of tricky. You want to secure three double cross blocks and two 9M beams with a pair of 8M cross axles with end stops (Figure 8-13).

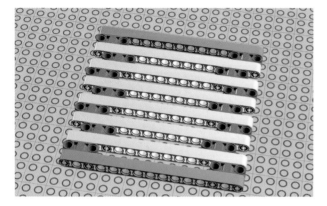

Figure 8-12. *Step 7: Two more 15M beams*

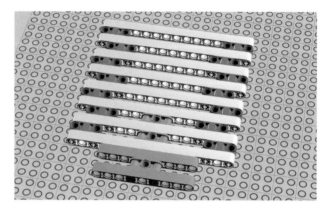

Figure 8-13. *Step 8: Adding two 9M beams and three double cross blocks*

9. Add another 15M beam and secure it with two 3M connector pegs as you see in Figure 8-14).

10. Add eight connector pegs as you see in Figure 8-15.

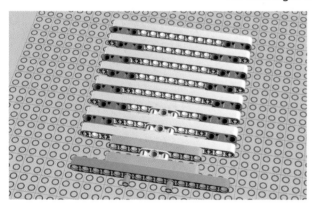

Figure 8-14. *Step 9: Securing a 15M beam with two 3M connector pegs*

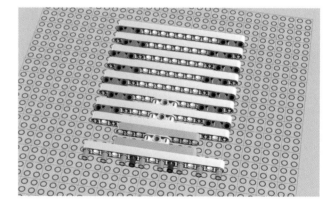

Figure 8-15. *Step 10: Adding eight additional pegs*

11. Connect two 14M Technic bricks to the pegs you just added (Figure 8-16).

12. Add two connector pegs to each Technic brick as seen in Figure 8-17.

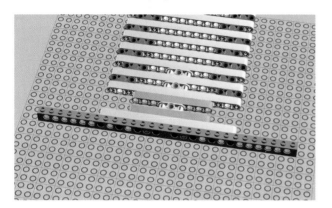

Figure 8-16. *Step 11: Connecting some bricks*

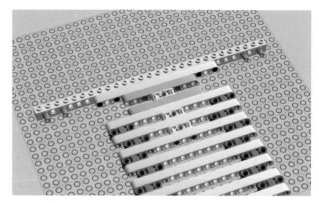

Figure 8-17. *Step 12: Attaching the pegs*

13. Add two 4x4 Technic bricks to the pegs (Figure 8-18).

14. Insert two connector pegs into the sides of the 4x4 Technic bricks (Figure 8-19).

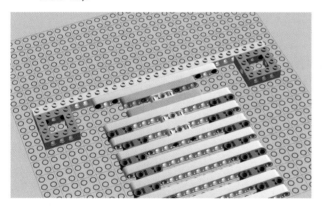

Figure 8-18. *Step 13: Connecting two 4x4 Technic bricks*

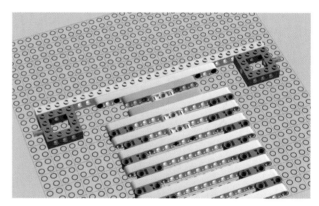

Figure 8-19. *Step 14: Inserting two connector pegs*

15. Add two 14M Technic bricks. Note that they will not want to stay still; we'll secure them in a future step (Figure 8-20).

16. Insert four connector pegs as shown in Figure 8-21.

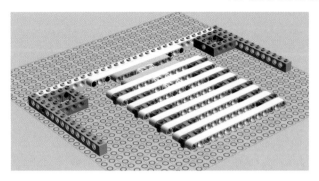

Figure 8-20. *Step 15: Connecting two more 14M Technic bricks*

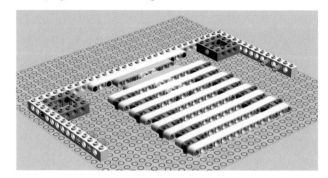

Figure 8-21. *Step 16: Attaching the pegs*

17. Add a pair of 9M beams (Figure 8-22).

18. Add five connector pegs, two in each 9M beam and one in the middle of the front 15M beam (Figure 8-23).

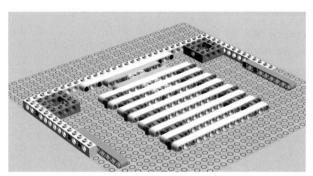

Figure 8-22. *Step 17: Adding the beams*

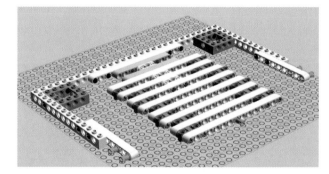

Figure 8-23. *Step 18: Connecting some more pegs*

19. Add a 4M Technic brick (Figure 8-24).

20. Secure another 4M Technic brick with two 3M connector pegs (Figure 8-25).

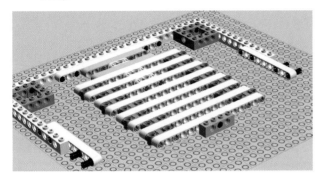

Figure 8-24. *Step 19: Connecting a 4M Technic brick*

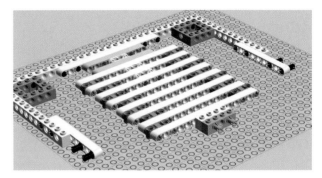

Figure 8-25. *Step 20: Adding a second brick with the help of two 3M pegs*

21. Add a third 4M Technic brick as seen in Figure 8-26.

22. Add a pair of 10M Technic bricks as seen in Figure 8-27.

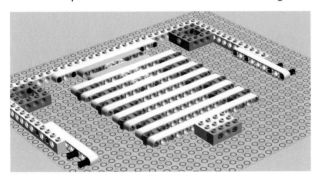

Figure 8-26. *Step 21: Adding another 4M Technic brick*

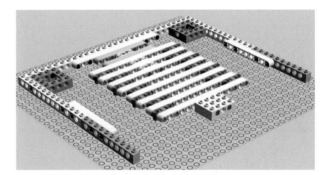

Figure 8-27. *Step 22: Adding two 10M Technic bricks*

23. Insert four connector pegs into the 10M Technic bricks and another into the trio of 4M bricks (Figure 8-28).

24. Connect two more 4x4 Technic bricks as seen in Figure 8-29.

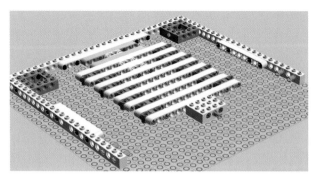

Figure 8-28. *Step 23: Adding five more connector pegs*

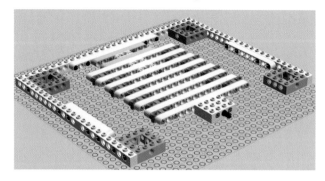

Figure 8-29. *Step 24: Adding two 4x4 Technic bricks*

25. Connect a 15M Technic beam and six connector pegs as shown in Figure 8-30.

26. Add a pair of 14M Technic bricks (Figure 8-31).

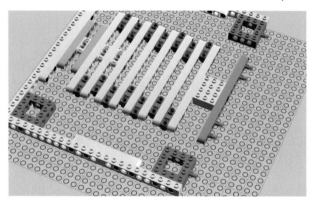

Figure 8-30. *Step 25: Adding a 15M beam and pegs*

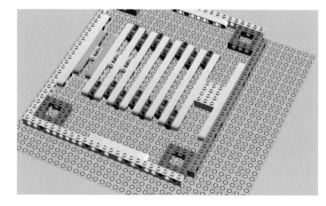

Figure 8-31. *Step 26: Connecting a pair of 14M Technic bricks*

Chapter 8

27. Let's add some plates! Add a left wing, a right wing, a 6x14 plate, and a 4x6 plate as seen in Figure 8-32.

28. Add nine connector pegs and three cross connectors to the 3M beams with pegs (Figure 8-33).

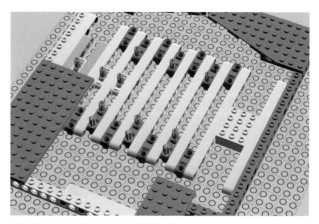

Figure 8-32. *Step 27: Reinforcing the Keytar body with a variety of Lego plates*

Figure 8-33. *Step 28: Adding a bunch of pegs*

29. Reinforce the Keytar body with a pair of 9M beams and a 7M beam (Figure 8-34).

30. Add seven more connector pegs (Figure 8-35).

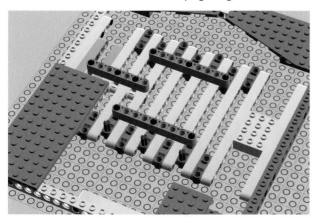

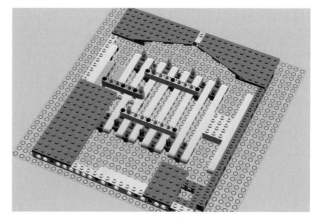

Figure 8-34. *Step 29: Adding two 9M beams and a 7M beam*

Figure 8-35. *Step 30: Adding more connector pegs*

31. Add a 9M beam and a 13M beam (Figure 8-36).

32. Reinforce the back with a layer of plates—count the studs in Figure 8-37 to get the sizes right!

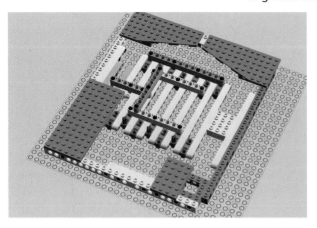

Figure 8-36. *Step 31: Adding a 9M beam and a 13M beam*

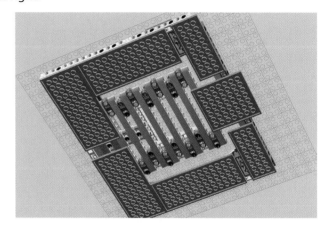

Figure 8-37. *Step 32: Reinforcing the Keytar body with a bevy of Lego plates*

33. Add another layer of plates as shown in Figure 8-38. The body is starting to come together!

34. Add a pair of 16M Technic bricks and four 14M Technic bricks to the underside of the Keytar body as shown in Figure 8-39.

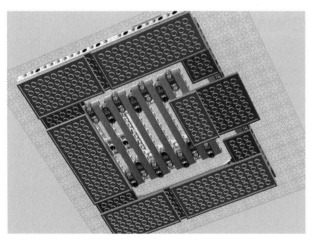

Figure 8-38. *Step 33: Adding more plates*

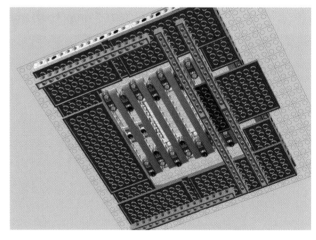

Figure 8-39. *Step 34: Adding additional reinforcement to the Keytar body*

35. Insert a two 3M connector pegs as seen in Figure 8-40.

36. Connect a 6x8 Technic brick to the pegs you just added in Step 35. You'll have to temporarily loosen the plate in order to add the 6x8 brick (Figure 8-41).

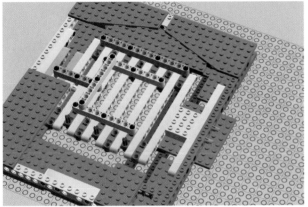

Figure 8-40. *Step 35: Adding a pair of 3M connector pegs*

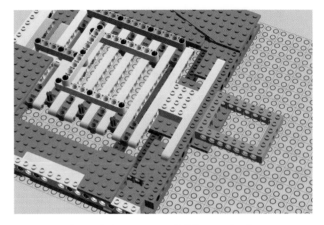

Figure 8-41. *Step 36: Connecting a 6x8 Technic brick*

37. Add four connector pegs to the 6x8 Technic brick (Figure 8-42).

38. Connect two 16M Technic bricks to the 6x8 brick, as shown in Figure 8-43.

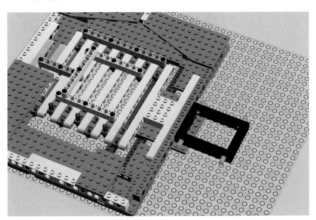

Figure 8-42. *Step 37: Adding four more pegs*

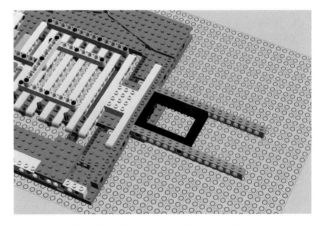

Figure 8-43. *Step 38: Adding a pair of Technic bricks*

39. Insert 10 more pegs (Figure 8-44).

40. Add two 5M beams as seen in Figure 8-45. The motors will mount to these beams later on.

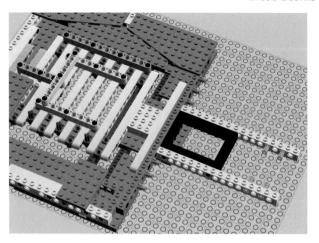

Figure 8-44. *Step 39: Inserting 10 more pegs*

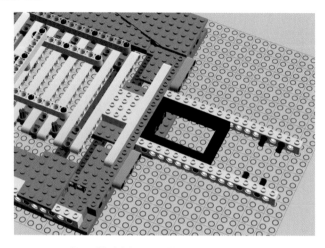

Figure 8-45. *Step 40: Adding two 5M Technic beams*

41. Add 9M beams as shown in Figure 8-46. The Keytar's neck is starting to take shape!

42. Insert four more pegs (Figure 8-47).

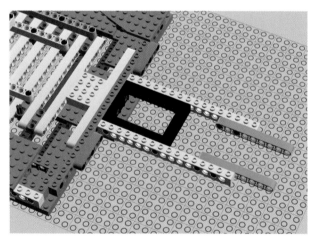

Figure 8-46. *Step 41: Adding a pair of 9M beams*

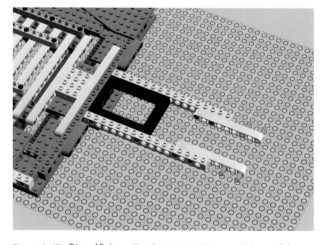

Figure 8-47. *Step 42: Inserting two connector pegs into each beam*

43. Add two 16M Technic bricks (Figure 8-48).

44. Reinforce the bottom of the neck with a pair of 8x8 plates (Figure 8-49).

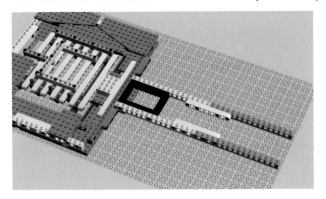

Figure 8-48. *Step 43: Adding 16M beams*

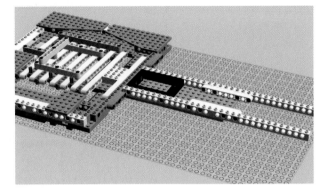

Figure 8-49. *Step 44: Adding a pair of 8x8 plates to the underside*

45. Add 6x8 and 6x10 plates to the top (Figure 8-50).

46. Add two 16M Technic bricks to the top (Figure 8-51).

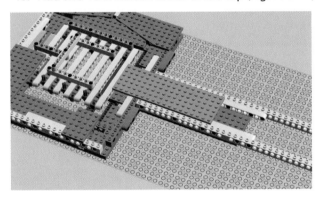

Figure 8-50. *Step 45: Adding 6x8 and 6x10 plates to the neck*

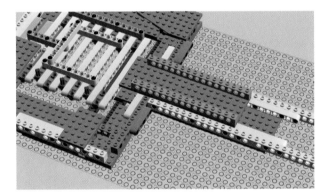

Figure 8-51. *Step 46: Adding a pair of bricks*

47. Add 6x14 and 8x8 plates to the underside (Figure 8-52).

48. Add two each 16M and 14M Technic bricks to the underside, as shown in Figure 8-53.

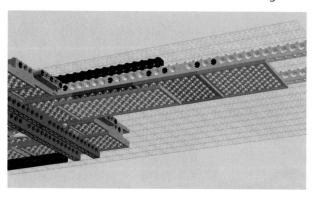

Figure 8-52. *Step 47: Adding more plates to the underside*

Figure 8-53. *Step 48: Attaching more Technic beams*

49. Secure a pair of 15M Technic beams with four 3M connectors (Figure 8-54).

50. Add two more 15M beams (Figure 8-55).

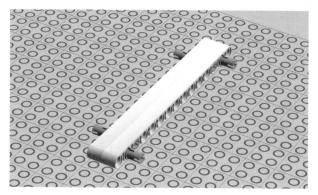

Figure 8-54. *Step 49: Combining connectors and beams*

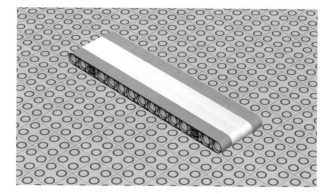

Figure 8-55. *Step 50: Adding two more 15M beams*

51. Set the four 15M beams on the neck as shown in Figure 8-56. Nope, they won't want to stay there (Figure 8-56).

52. Secure the Technic beam assembly with a pair of 10M cross axles and bushes (Figure 8-57).

Figure 8-56. *Step 51: Setting the beam assembly on the neck*

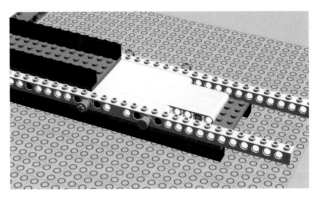

Figure 8-57. *Step 52: Securing the beam assembly with two 10M cross axles and bushes*

53. Add eight more connector pegs as shown in Figure 8-58.

54. Attach two 3M beams and two 9M beams (Figure 8-59).

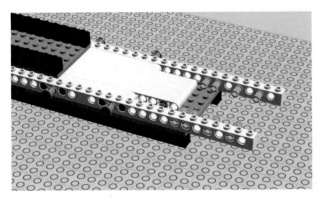

Figure 8-58. *Step 53: Adding eight more pegs*

Figure 8-59. *Step 54: Attaching two each 3M and 9M beams*

55. Insert four more pegs as shown in Figure 8-60.

56. Add two 14M Technic bricks (Figure 8-61).

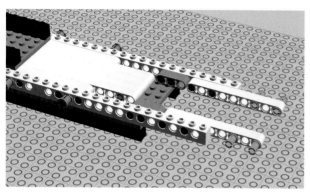

Figure 8-60. *Step 55: Adding four connector pegs*

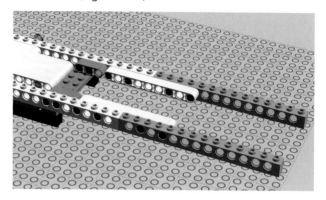

Figure 8-61. *Step 56: Adding a pair of 14M Technic bricks*

57. Adding two each 10M and 8M Technic bricks to the underside (Figure 8-62).

58. Add eight more connector pegs (Figure 8-63).

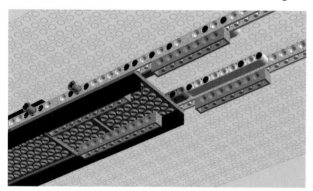

Figure 8-62. *Step 57: Reinforcing the bottom with four more Technic bricks*

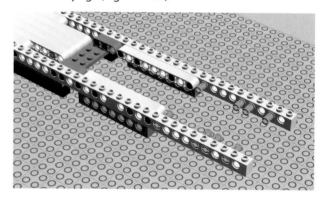

Figure 8-63. *Step 58: Adding eight pegs*

59. Add two each 5M and 9M Technic beams (Figure 8-64).

60. Insert four more pegs (Figure 8-65).

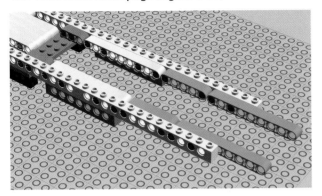

Figure 8-64. *Step 59: Adding two each 5M and 9M beams*

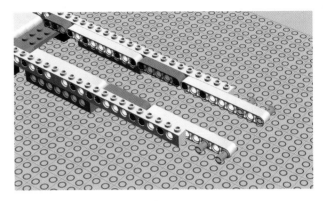

Figure 8-65. *Step 60: Inserting four more pegs*

61. Connect two 14M Technic bricks as shown in Figure 8-66.

62. Insert four more connector pegs (Figure 8-67).

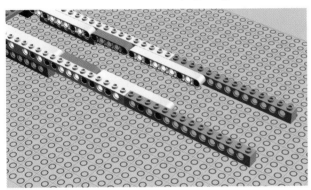

Figure 8-66. *Step 61: Adding 14M bricks*

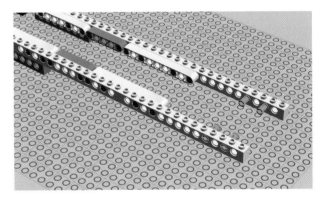

Figure 8-67. *Step 62: Inserting more pegs*

63. Add two more 9M beams (Figure 8-68).

64. Insert a pair of pegs as seen in Figure 8-69.

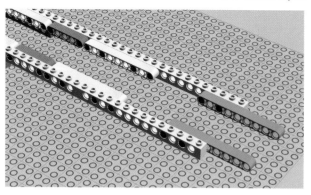

Figure 8-68. *Step 63: Adding a pair of 9M beams*

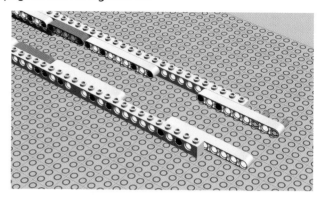

Figure 8-69. *Step 64: Adding more pegs*

65. Add a 14M Technic brick (Figure 8-70).

66. Add two more connector pegs. Are you getting tired of pegs yet (Figure 8-71)?

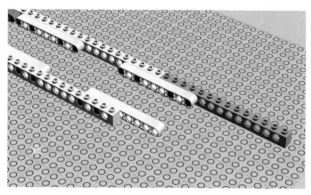

Figure 8-70. *Step 65: Adding a 14M beam*

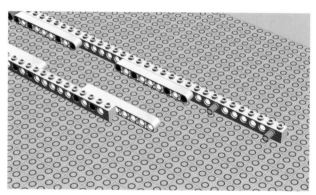

Figure 8-71. *Step 66: Inserting more pegs*

67. Connect a 6x8 Technic brick as shown in Figure 8-72.

68. Add four connector pegs. The neck is nearly done (Figure 8-73)!

Figure 8-72. *Step 67: Connecting a brick*

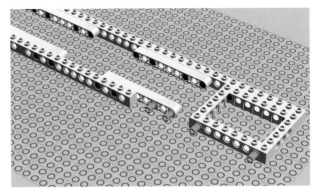

Figure 8-73. *Step 68: Adding even more pegs*

69. Attach a 14M Technic brick as shown in Figure 8-74.

70. Add a 6x8 plate to the end of the neck as seen in Figure 8-75.

Figure 8-74. *Step 69: Connecting a 14M brick*

Figure 8-75. *Step 70: Adding a 6x8 plate*

71. Add four more plates to the underside: an 8x8 plate at the very end, and 4x8s per Figure 8-76.

72. Add four 4M and two 6M Technic bricks (Figure 8-77).

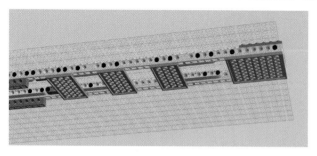

Figure 8-76. *Step 71: Adding plates to the underside of the neck*

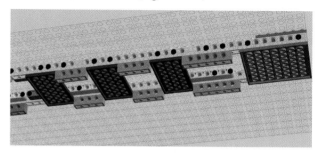

Figure 8-77. *Step 72: Adding more bricks to the underside*

73. Insert 20 cross connectors as shown in Figure 8-78.

74. Connect cross blocks to the pegs you just added (Figure 8-79).

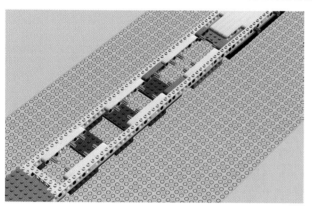

Figure 8-78. *Step 73: Adding a herd of cross connectors*

Figure 8-79. *Step 74: Adding cross blocks*

75. Then, add even more cross connectors! The button plates will mount on these pegs (Figure 8-80).

76. Add four connector pegs as shown in Figure 8-81. The motors will mount to these pegs.

Figure 8-80. *Step 75: Inserting another 20 cross connectors*

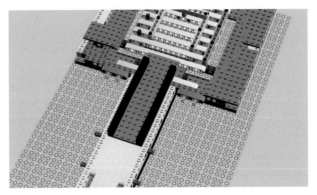

Figure 8-81. *Step 76: Inserting connector pegs*

77. Connect the motors (Figure 8-82).

78. Add four 9M cross axles and a pair of cross axle connectors (Figure 8-83).

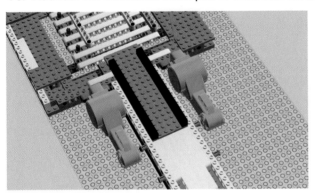

Figure 8-82. *Step 77: Adding motors*

Figure 8-83. *Step 78: Adding 9M axles and connectors*

79. Secure the ends of the cross axles with bushes (Figure 8-84).

80. Insert pole reversal handles into the motors' hubs (Figure 8-85).

Figure 8-84. *Step 79: Adding bushes*

Figure 8-85. *Step 80: Adding pole reversal handles to the motors' hubs*

81. Slide 8M cross axles with end stops through the pole reversal handles as seen in Figure 8-86.

82. Slide tubes onto the cross axles (Figure 8-87).

Figure 8-86. *Step 81: Attaching 8M cross axles with end stops*

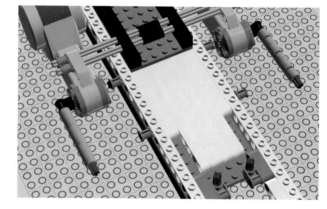

Figure 8-87. *Step 82: Adding tubes to the cross axles*

83. Secure the ends of the cross axles with ball elements (Figure 8-88).

84. You're finished! Your Keytar should look like Figure 8-89.

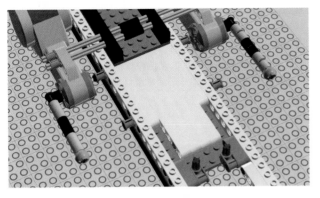

Figure 8-88. *Step 83: Securing the ends of the cross axles with ball elements*

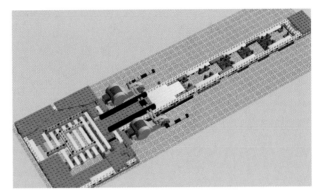

Figure 8-89. *Step 84: You're done!*

Install the Electronics

When you've built the Keytar, then it's time to add the electronics. Follow these easy steps to complete the project:

- Attach the Arduino to its mounting plate and seat the Bricktronics shield on the Arduino.
- Connect the mounting plate to the Keytar and secure with half-bushes.
- Add the buttons to the mounting plates and wire up the buttons' leads.
- Slide the wires through the holes in the Keytar's neck, then secure the mounting plates to the neck with half-bushes. You can see a close-up of this in Figure 8-2.
- Connect the button wires to pins 6, 7, and A1-A3 as seen in Figure 8-90.
- Add Mindstorms wires for the touch sensor (the mute button) and the two motors.
- Connect the piezo as indicated in Figure 8-90.

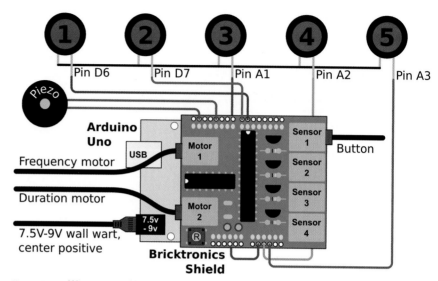

Figure 8-30. *Wire up your Keytar as you see here*

Program the Keytar

Download the following code to the Arduino to make the Keytar work:

```
#include <Wire.h>
#include <Adafruit_MCP23017.h>
#include <Bricktronics.h>

// Make: Lego and Arduino Projects
// Chapter 8: Keytar
// Website: http://www.wayneandlayne.com/bricktronics/

Bricktronics brick = Bricktronics();
Motor tone_knob = Motor(&brick, 1);
Motor duration_knob = Motor(&brick, 2);
Button mute_button = Button(&brick, 1);

#define SPEAKER 11 ❶
#define TONE_SLOT_0 6
#define TONE_SLOT_1 7
#define TONE_SLOT_2 A1
#define TONE_SLOT_3 A2
#define TONE_SLOT_4 A3
#define NUM_OF_SLOTS 5

char slot_button[] = { ❷
    TONE_SLOT_0,
    TONE_SLOT_1,
    TONE_SLOT_2,
    TONE_SLOT_3,
    TONE_SLOT_4};

int slot_frequency[] = {300, 400, 500, 600,700}; ❸
int note_duration = 500; ❹
int rest_duration = 100; ❺
int current_slot = 0; ❻
```

```
#define MIN_FREQUENCY 25  ❼
#define MIN_NOTE_DURATION 25

void setup()  ❽
{
    Serial.begin(115200);  ❾
    Serial.println("starting!");
    brick.begin();
    tone_knob.begin();
    duration_knob.begin();
    mute_button.begin();

    for (char i = 0; i < NUM_OF_SLOTS; i++)  ❿
    {
        pinMode (slot_button[i], INPUT_PULLUP);
    }

    while (!mute_button.is_pressed()) {
        //do nothing
    }  ⓫
}

void loop()  ⓬
{
    Serial.println(current_slot);
    int current_frequency = slot_frequency[current_slot];  ⓭

    if (!mute_button.is_pressed())
    {
        tone(SPEAKER, current_frequency);  ⓮
    }

    watch_for_input(note_duration, current_slot, false, true);  ⓯
    noTone(SPEAKER);
    watch_for_input(rest_duration, current_slot, false, false);  ⓰

    current_slot += 1;  ⓱

    if (current_slot == NUM_OF_SLOTS)
    {
        current_slot = 0;
    }
}


void watch_for_input(long ms, char slot, boolean in_slot, \
    boolean is_playing)  ⓲
{
    long start_time = millis();
    long end_time = start_time + ms;
    tone_knob.encoder->write(0);  ⓳
    duration_knob.encoder->write(0);
    Serial.print("In slot: ");    Serial.println(in_slot, DEC);
    while (millis() < end_time)  ⓴
    {
        if (!in_slot)    ㉑
        {
            for (char i = 0; i < NUM_OF_SLOTS; i++)
            {
                if (digitalRead(slot_button[i]) == LOW)
```

```
                            {
                                Serial.print("Button pressed: ");
                                Serial.println(i, DEC);
                                handle_button_press(i);
                            }
                        }
                    }

            long duration_change = duration_knob.encoder->read();

            if (duration_change != 0)
            {
                Serial.print("duration knob: ");
                Serial.println(duration_change, DEC);
                if (in_slot)  ❷❷
                {
                    note_duration += duration_change;
                    note_duration = max(note_duration, MIN_NOTE_DURATION);  ❷❸
                }
                else
                {
                    rest_duration += duration_change;
                    rest_duration = max(rest_duration, 0);
                }
                end_time = millis();      ❷❹
                duration_knob.encoder->write(0);  ❷❺
            }

            if (tone_knob.encoder->read() != 0)
            {
                Serial.print("Tone knob: ");
                Serial.println(tone_knob.encoder->read(), DEC);

                slot_frequency[slot] += (tone_knob.encoder->read());
                slot_frequency[slot] = max(MIN_FREQUENCY, slot_frequency[slot]);
                if (is_playing && !mute_button.is_pressed())
                {
                    tone(SPEAKER, slot_frequency[slot]);
                }
                tone_knob.encoder->write(0);
            }
        }
    }

void handle_button_press(char slot)  ❷❻
{
    Serial.print("Handling button press for slot: ");
    Serial.println(slot);
    noTone(SPEAKER);
    while (digitalRead(slot_button[slot]) == LOW)
    {
        tone(SPEAKER, slot_frequency[slot]);
        watch_for_input(note_duration, slot, true, true);
        noTone(SPEAKER);
        watch_for_input(rest_duration, slot, true, false);
    }
}
```

❶ These `defines` set up labels for the pins used in this sketch.

❷ `slot_button` is an array. Conceptually, each button corresponds to a slot. Each slot has a frequency. This array maps slot numbers to the pin that corresponds to that slot's button.

❸ `note_duration` holds the duration of the note in milliseconds. As you play, you'll change this from the initial value.

❹ `slot_frequency[]` is an array of `ints` that contain the frequencies for each slot. It starts with the initial frequencies for each slot.

❺ `rest_duration` holds the duration in milliseconds of the rest between tones. As you play, you'll change this from the initial value.

❻ `current_slot` is a global variable that holds the current slot. As the notes progress, this will increase until it reaches the end of the slots, at which point it will wrap around back to zero.

❼ `MIN_FREQUENCY` is a `define` that sets the minimum frequency for any slot.

❽ The `setup()` function is called only once—right after the Arduino is powered on.

❾ We use the USB serial port for debugging our Arduino code. It allows us to send information from the Arduino to the computer.

❿ This `for` loop initializes each of the slot buttons as an input, with internal pullups turned on.

⓫ This `while` loop stops the program from proceeding at startup until the mute button is pressed.

⓬ The `loop()` function runs over and over until you remove power from the Arduino.

⓭ `current_frequency` is set to the integer stored in the `current_slot` index of the array `slot_frequency`.

⓮ This starts a tone on pin `SPEAKER` with frequency `current_frequency`.

⓯ This calls the function that waits for `note_duration` milliseconds and responds to any input during that time.

⓰ The details of this function will be presented described shortly; it waits for `rest_duration` milliseconds and responds to any input during that time.

⓱ The end of this function sets `current_slot` to the next slot, in preparation for the next run of `loop()`.

⓲ `watch_for_input` responds to input for a specific amount of time. It has four input parameters:

> *ms*
> How long to watch for input in milliseconds.

> *slot*
> The current slot that any input will modify.

in_slot

A boolean that indicates if a slot fret button is pressed.

is_playing

This indicates whether the note in that slot is actually being played.

⑲ By writing a zero to the encoder object, it is easy to see later if the motors have moved. If a later call to read doesn't return 0, they've moved!

⑳ This while loop keeps checking for input until the current time is the end_time.

㉑ If no fret buttons are pressed, we check every fret button to see if it is newly pushed.

㉒ If the duration knob is turned, and a fret button is pushed, change the note time. Otherwise, change the rest time.

㉓ This limits the note duration to a minimum of MIN_NOTE_DURATION.

㉔ This statement makes the function exit at the end of this iteration of the while loop. If the knobs are spinning, this means they'll likely still be spinning a bit at the next slot, which makes an awesome retro arcade-style chunky sound.

㉕ Now that we've handled the duration knob, we set it to 0.

㉖ handle_button_press handles fret button presses. It contains the main logic for "fret button mode."

Play Some Music!

When you turn on the Keytar (Figure 8-91), nothing happens until the mute button is pressed once. If nothing is pressed after that, the Keytar plays one note and one rest for each of its five slots, and then loops back to the beginning. If the duration knob is turned, it changes the duration of the rests in between the notes. If a "fret" button is pressed, the note corresponding to that slot plays, along with a rest, over and over. If the duration knob is turned while the fret button is pressed, it changes the duration of all the notes. If the tone knob is turned, regardless of a fret button being pressed, it changes the frequency of the current slot. You can make awesome retro arcade noises by turning the frequency knob with none of the fret buttons pressed! At any time, pressing the mute button stops the next note from being played. When it's released, the next note will be played.

Figure 8-91. *You're done! It's time to scare the cat.*

The Next Chapter

Our next chapter offers one last project to hone your Mindstorms and Arduino skills to perfection. In Chapter Nine we'll show you how to make a Lego lamp that can be turned on and off or dimmed with a smartphone application!

Project: Lamp

Figure 9-1. *The Lamp project creates a Lego light that can be controlled via smartphone*

For our final project we're going to create a lamp out of Legos that uses translucent bricks as the shade. We chose a classic kid's night light for the bulb, in part because it's very compact—all the better to fit in the modest confines of the Lamp project's interior—as well as because of its low heat output.

You might express alarm at the prospect of triggering wall current with your Arduino. Is it safe? As long as you're careful! We use a hobbyist power cord (see the sidebar, "PowerSwitch Tails") that connects the night light to AC current in a way that's as safe and effective as flicking a light switch. It's a great tool that will encourage you to do more high voltage projects without your Arduino without getting fried.

For fun, we'll add a Bluetooth Shield, so we can control the lamp remotely. We've written an Android app that works with most Android phones and tablets. It's available on the Google Play store, as well as on the Bricktronics website. There have been some reports that the Bluetooth Shield we use in this project doesn't work with a few Android phones—check the Bricktronics page for compatibility information. Once you have it installed, turn Bluetooth on in your Android device's settings, and then pair the device with the nightlight.

Parts List

Let's begin the Lamp project by gathering together all the parts you need. One of the most surprising may be a phone (Figure 9-2) that we use to control the lamp via Bluetooth. However, if you just want to control the lamp manually, that works too!

Tools & Electronics

You'll need the following non-Lego parts to build your lamp:

- Arduino Uno

- Bricktronics Shield

- Mounting plate for Arduino

- Power supply rated for 9V at 1.3A or greater with a 2.1mm center-positive plug. This provides power to the Arduino.

- Bluetooth Shield (Seeedstudio PN SLD63030P)

- PowerSwitch Tails. We used the PowerSSR Tail and the ZeroCross Tail, available at *http://powerswitchtail.com*.

- Night light (any standard incandescent light found in hardware and grocery stores will be fine)

- Zip ties

- Wire management wraps like the Vaisis Spiral Harness Wrap. Do a search on Amazon to find the diameter you want.

- Android phone (optional)

- 22-gauge solid-core wires (we used Maker Shed PN MKEE3)

- 1K resistor

Figure 9-2. *In this project you'll get to control Lego with an Android smartphone!*

POWERSWITCH TAILS

Most people would prefer not exposing themselves to high voltage electrical sources, which is why we chose Power-Switch Tails (Figure 9-3) to power our lamp. They're power cords easily triggered by microcontrollers, allowing you to start and stop the flow of electricity with a program. They plug into any standard three-prong outlet, extension cord, or power strip.

What all can you do with these cords? You could easily make a motion-activated work light, for instance. Or how about a fan triggered by a temperature sensor, so the air kicks in whenever it gets hot? Of course, you could always buy equivalent products, but what fun is that? The bottom line is that these products let hobbyists safely trigger wall current.

Now, you might ask, why are we using *two* PowerSwitch Tails? The answer is you need two in order to provide dimming functionality, because a regular PowerSwitch Tail turns power on and off, but can't send just a trickle of electricity to dim the light.

This is how it works: The power from the wall sockets is alternating current, or AC. The voltage on an AC line can be represented with a sine wave. It's positive for awhile, then zero, then negative for awhile. A good way to dim lights that plug into AC is with an electronic component called a TRIAC. A TRIAC can conduct electricity in both directions when it is turned on. Once it is turned on, it will stay on until the current drops below a threshold, called the *holding current*. If we turn the TRIAC on for every wave, and turn it on as soon

as possible, then the light connected to the end will be fully on. If we never turn the TRIAC on, the light connected to the end will be fully off. If we can turn the TRIAC on at the same point in the AC wave, we can get an intermediate brightness. The brightness will depend upon where in the wave we turn the TRIAC on.

Because we know how long a wave is for AC (mostly 60 Hz or 50 Hz depending upon where you live), as long as we start counting from a zero cross, we can use timing to set our brightness. The longer we wait after a zero cross to turn on, the dimmer it will be.

To do this without exposing the Arduino to AC, we use two different products—the ZeroCross Tail and the PowerSSR Tail. The ZeroCross Tail is a pass-through socket with a ground and a zero-cross output, which will go low every time there is a zero cross. The PowerSSR Tail has a TRIAC in it. If we connect it to the ZeroCross Tail and the TRIAC input to the Arduino, we can dim the night light inside the Lamp project. We can watch the zero-cross output, and once it goes low, we can start waiting, and turn on the TRIAC as soon as the appropriate amount of time goes by.

If you just want to turn the light on and off, you can buy the classic PowerSwitch Tail II at the Maker Shed: *http://www. makershed.com/PowerSwitch_Tail_II_p/mkps01.htm*. Alternatively, the PowerSSR Tail and the ZeroCross Tail used in this project are available at *http://powerswitchtail.com*.

Figure 9-3. *The PowerSwitch Tail is a convenient and safe way to control wall current*

Lego Elements

The Lamp project is unique in this book because it uses a lot of regular Lego parts (Figure 9-4) rather than focusing on Mindstorms. The following list describes all the Lego elements you'll need:

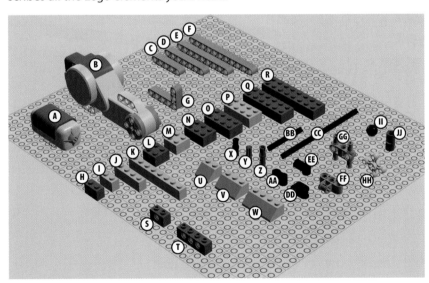

Figure 9-4. *You'll need a lot of bricks for this project!*

A. 1 Mindstorms touch sensor

B. 1 Mindstorms motor

C. 3 Technic beams 5M

D. 2 Technic beams 7M

E. 1 Technic beam 9M

F. 3 Technic beams 11M

G. 4 Technic 90-degree angle beams 3x5

H. 8 Lego bricks 1x2

I. 8 Lego translucent bricks 1x2

J. 2 Lego translucent bricks 1x4

K. 11 Lego translucent bricks 1x8

L. 30 Lego bricks 2x2

M. 20 Lego translucent bricks 2x2

N. 6 Lego bricks 2x3

O. 62 Lego bricks 2x4

P. 45 Lego translucent bricks 2x4

Q. 20 Lego bricks 2x6

R. 15 Lego bricks 2x8

S. 2 Technic bricks 1x2

T. 1 Technic bricks 1x8

U. 59 Lego translucent 45-deg slopes 2x2

V. 10 Lego translucent 45-deg slopes 2x3

W. 19 Lego translucent 45-deg slopes 2x4

X. 20 Technic pegs 2M

Y. 7 Technic cross connectors

Z. 5 Technic pegs 3M

AA. 2 Technic dampers

BB. 1 Technic cross axle 5M

CC. 1 Technic cross axle 12M

DD. 1 Technic cross block 1x2

EE. 1 Technic 0-degree angle element

FF. 1 Technic cross block 3x2

GG. 3 Technic beam 3M w/pegs

HH. 1 Technic angular wheel

II. 1 Technic ball element

JJ. 4 tubes

KK. 2 Mindstorms wires (not pictured)

Assembly Instructions

Let's get started building the Lamp project! Hopefully piling one classic brick on top of the next brings you back to the good ol' days of Lego building!

One thing to note about these step-by-steps is the color we use to connote translucent bricks. Lego Digital Designer, the tool we used for our model renderings, faithfully tries to depict what a translucent brick might look like, but that makes it kind of hard to distinguish the different layers of bricks. We chose to use a blue color to represent translucent clear bricks (Figure 9-5) so any time you see that color in this chapter, know that you won't actually be using opaque blue bricks—unless you want the dimmest lamp in history!

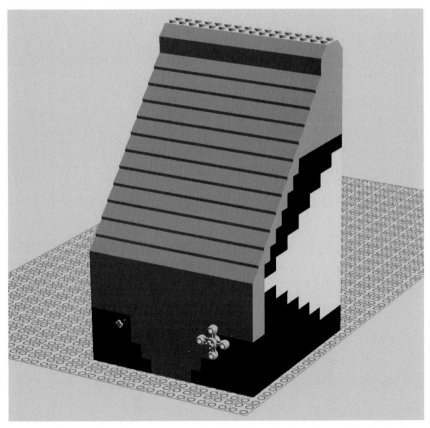

Figure 9-5. *Remember, wherever you see blue bricks, imagine translucent clear instead!*

Build the Lego Model

Let's get started building the Lamp project. It consists of a shell of classic Lego bricks, with Mindstorms guts managing the manual control of the on/off and dimming functionality.

1. Let's begin with two 2x2 Lego bricks and a 2x4 translucent brick (Figure 9-6).

2. Add two more 2x4 bricks, per Figure 9-7. Note that these aren't actually connected to each other, just sitting side-by-side.

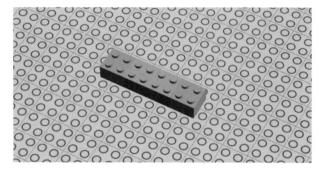

Figure 9-6. Step 1: Adding 2x2 bricks to a 2x4 translucent brick

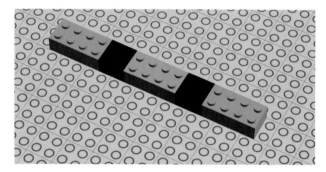

Figure 9-7. Step 2: Attaching two 2x4 bricks

3. Add two more 2x2 bricks (Figure 9-8).

4. Add six 2x4 bricks (Figure 9-9).

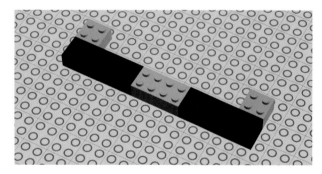

Figure 9-8. Step 3: Adding two 2x2 bricks

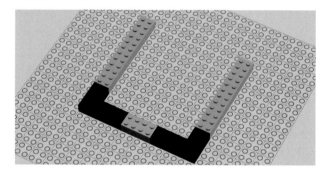

Figure 9-9. Step 4: Adding more 2x4 bricks

5. Next, attach three 2x2 translucent bricks (Figure 9-10).

6. Add a pair of 1x2 bricks (Figure 9-11).

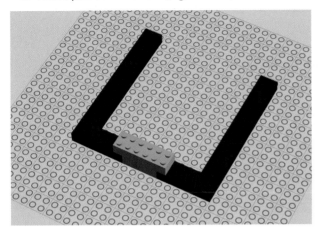

Figure 9-10. *Step 5: Adding three 2x2 translucent bricks*

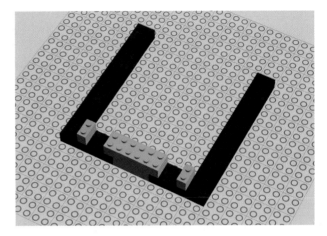

Figure 9-11. *Step 6: Attaching a pair of 1x2 bricks*

7. Add four more 2x2 bricks (Figure 9-12).

8. Finish off the second layer with 10 2x4 bricks (Figure 9-13).

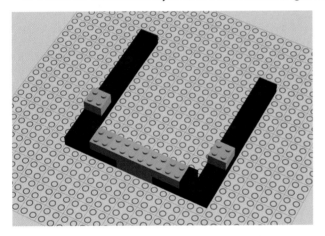

Figure 9-12. *Step 7: Adding four 2x2 bricks*

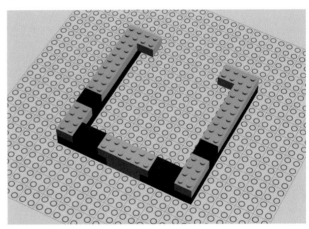

Figure 9-13. *Step 8: Placing eight 2x4 bricks*

9. Then add two more 2x4 bricks (Figure 9-14).

10. Add two 2x2 and one 2x4 translucent bricks (Figure 9-15).

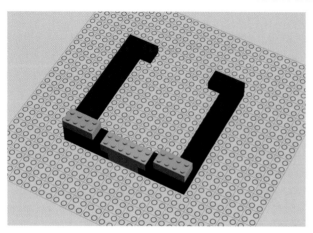

Figure 9-14. *Step 9: Adding more 2x4 bricks*

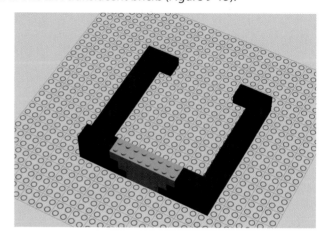

Figure 9-15. *Step 10: Adding two 2x2 and one 2x4 translucent bricks*

11. Next, add two 1x1 Technic bricks as seen in Figure 9-16.

12. Now you need two 1x2 bricks (Figure 9-17).

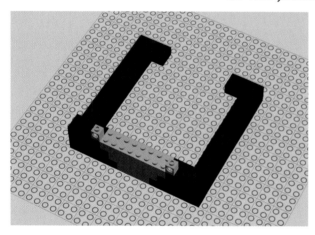

Figure 9-16. *Step 11: Attaching a couple of 1x1 Technic bricks*

Figure 9-17. *Step 12: Two more bricks*

13. Add four 2x2 bricks (Figure 9-18).

14. Connect five 2x4 bricks (Figure 9-19).

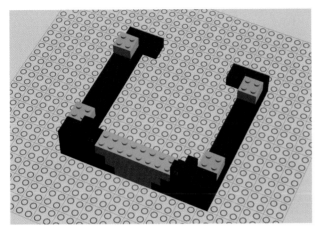

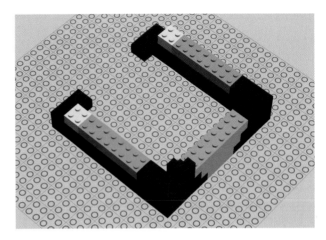

Figure 9-18. *Step 13: Adding four 2x2 bricks*

Figure 9-19. *Step 14: Connecting four 2x4 bricks*

15. Add four 2x6 bricks (Figure 9-20).

16. Add three 2x2 and two 2x1 translucent bricks (Figure 9-21).

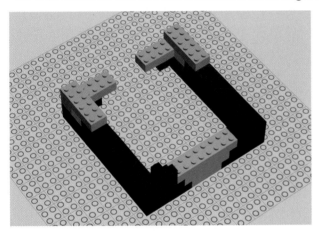

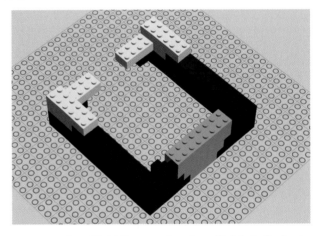

Figure 9-20. *Step 15: Adding four 2x6 bricks*

Figure 9-21. *Step 16: Adding more translucent bricks to the front*

17. Now you need four 2x2 bricks (Figure 9-22).

18. Next, add four 2x4 bricks (Figure 9-23).

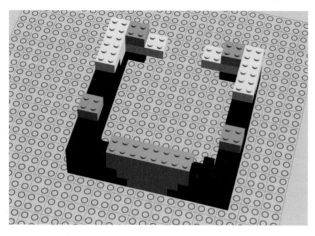

Figure 9-22. *Step 17: Adding four 2x2 bricks*

Figure 9-23. *Step 18: Placing four 2x4 bricks*

19. Connect three 2x8 bricks. Trap a 2x4 brick underneath the middle one (Figure 9-24).

20. Add two 2x2 and three 2x4 translucent bricks (Figure 9-25).

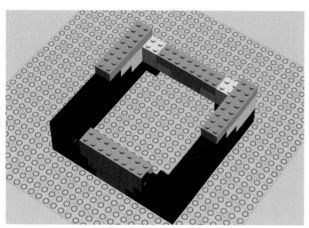

Figure 9-24. *Step 19: Adding three 2x8 and one 2x4 bricks*

Figure 9-25. *Step 20: Attaching a whole row of translucent bricks*

21. Add two 2x2 and two 2x6 bricks (Figure 9-26).

22. Add four 2x4 translucent bricks as shown in Figure 9-27.

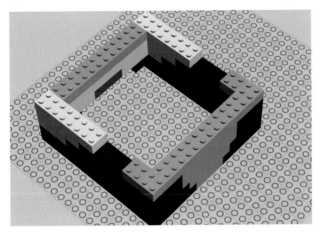

Figure 9-26. *Step 21: More bricks*

Figure 9-27. *Step 22: Placing another row of translucent bricks*

23. Next, you need three 2x4 bricks (Figure 9-28).

24. Add two 2x6 bricks (Figure 9-29).

Figure 9-28. *Step 23: Adding three 2x4 bricks*

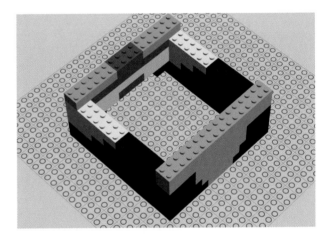

Figure 9-29. *Step 24: Adding two 2x6 bricks*

25. Add two 2x8 bricks (Figure 9-30).

26. Time to another row of translucent bricks. Add two 2x2s and three 2x4s (Figure 9-31).

Figure 9-30. *Step 25: Adding two 2x8 bricks*

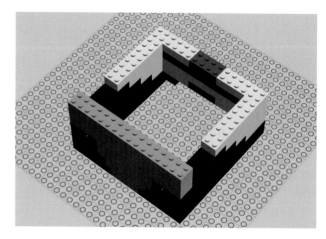

Figure 9-31. *Step 26: Adding another row of translucent bricks*

27. Add four 2x4 bricks (Figure 9-32).

28. Add a 2x8 brick (Figure 9-33).

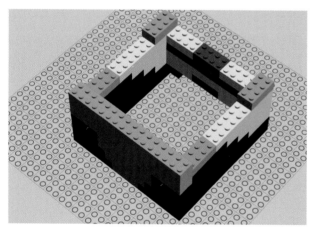

Figure 9-32. *Step 27: Adding four 2x4 bricks*

Figure 9-33. *Step 28: Adding a 2x8 brick*

29. Attach four 2x2 bricks (Figure 9-34).

30. Add two 2x3 bricks (Figure 9-35).

Figure 9-34. *Step 29: Placing four 2x2 bricks*

Figure 9-35. *Step 30: Adding a pair of 2x3 bricks*

31. Add four 2x8 bricks. These monsters are great for reinforcing your model (Figure 9-36)!

32. Add three 2x6 bricks as shown in Figure 9-37.

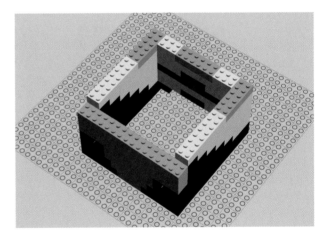

Figure 9-36. *Step 31: Reinforcing things with four 2x8 bricks*

Figure 9-37. *Step 32: Adding three 2x6 bricks*

33. Now you need four 2x4 translucent bricks (Figure 9-38).

34. Add an 8M Technic brick to the back. Your Arduino will mount to this brick (Figure 9-39).

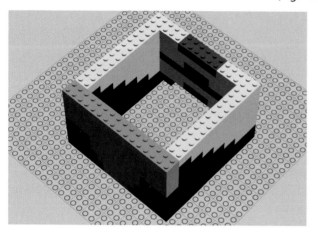

Figure 9-38. *Step 33: Adding four 2x4 translucent bricks*

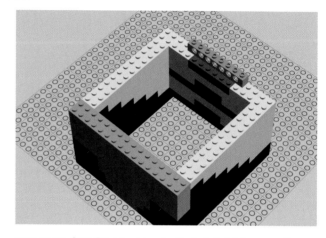

Figure 9-39. *Step 34: Attaching the Technic brick for mounting the Arduino*

35. Add two 2x4 bricks (Figure 9-40).

36. Next, add two 2x8 bricks (Figure 9-41).

Figure 9-40. *Step 35: Adding a pair of 2x4 bricks*

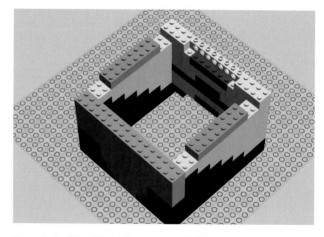

Figure 9-41. *Step 36: Adding a pair of 2x8 bricks*

37. Add two 2x2 bricks (Figure 9-42).

38. Now we get to work on the sloping part of the lamp. Add two 2x2 and three 2x4 translucent slopes (Figure 9-43).

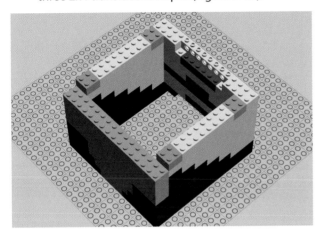

Figure 9-42. *Step 37: Adding 2x2 bricks*

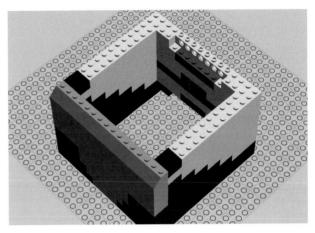

Figure 9-43. *Step 38: Adding translucent slopes*

39. Add two 2x3 and two 2x4 bricks (Figure 9-44).

40. Add two 2x3 and three 2x6 bricks (Figure 9-45).

Figure 9-44. *Step 39: Attaching two 2x3 and two 2x4 bricks*

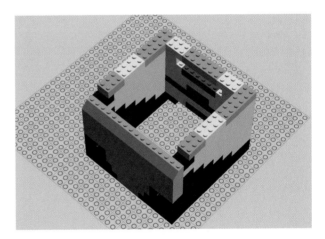

Figure 9-45. *Step 40: Adding two 2x3 and three 2x6 bricks*

41. Continue with the translucent slopes with two 2x2s and three 2x4s (Figure 9-46).

42. Add two 2x4 and two 2x6 bricks (Figure 9-47).

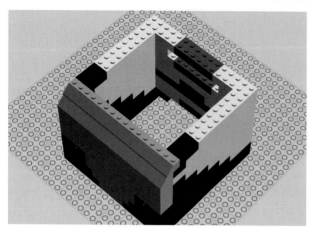

Figure 9-46. *Step 41: Attaching another row of slopes*

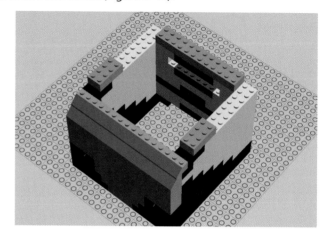

Figure 9-47. *Step 42: Adding two 2x4 and two 2x6 bricks*

43. Add two 2x8 bricks (Figure 9-48).

44. Now it's time for another row of translucent slopes. We used five 2x2s and two 2x3s (Figure 9-49).

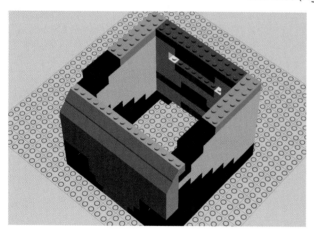

Figure 9-48. *Step 43: Adding two 2x8 bricks*

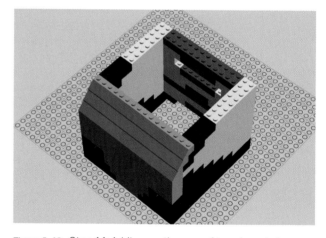

Figure 9-49. *Step 44: Adding another row of translucent slopes*

45. Add five 2x4 bricks (Figure 9-50).

46. Add two 2x1 and two 2x6 bricks (Figure 9-51).

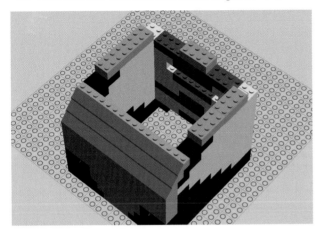

Figure 9-50. *Step 45: Adding five 2x4s*

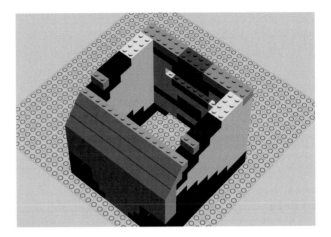

Figure 9-51. *Step 46: Adding two 2x1 and two 2x6 bricks*

47. Next, add two 2x2 and three 2x4 translucent slopes (Figure 9-52).

48. Add four 2x4 bricks (Figure 9-53).

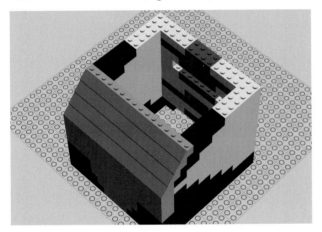

Figure 9-52. *Step 47: Adding more translucent slopes*

Figure 9-53. *Step 48: Adding four 2x4 bricks*

49. Add two 2x2 and two 2x6 bricks (Figure 9-54).

50. Now you need four 2x2 and two 2x4 translucent slopes (Figure 9-55).

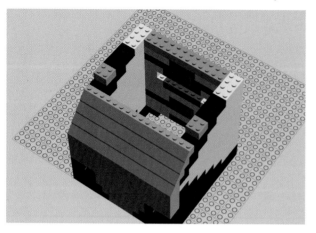

Figure 9-54. *Step 49: Adding two 2x2 and two 2x6 bricks*

Figure 9-55. *Step 50: Adding even more translucent slopes*

51. Add four 2x4 bricks (Figure 9-56).

52. Add two 2x1 bricks (Figure 9-57).

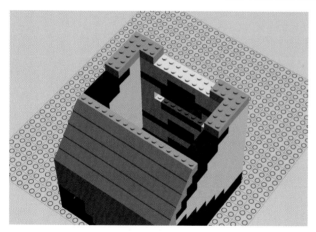

Figure 9-56. *Step 51: Adding four 2x4 bricks*

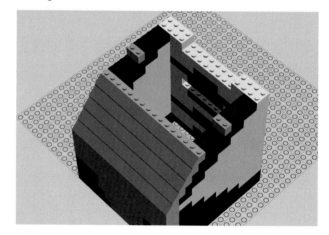

Figure 9-57. *Step 52: Adding a pair of 2x1 bricks*

Chapter 9

53. Add two 2x2 and one 2x8 bricks (Figure 9-58).

54. Next, add two 2x3, two 2x4, and one 2x2 translucent slopes (Figure 9-59).

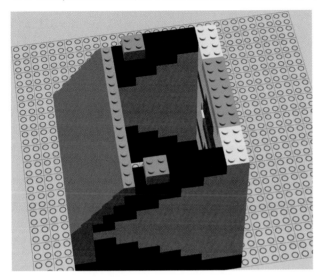

Figure 9-58. *Step 53: Adding two 2x2 and one 2x8 bricks*

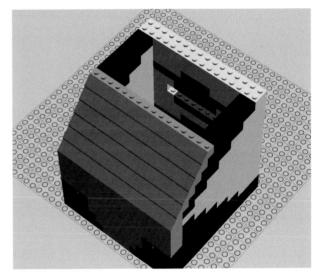

Figure 9-59. *Step 54: Attaching more slopes*

55. Add seven 2x4 bricks (Figure 9-60).

56. Time for six 2x2 and one 2x4 translucent slopes (Figure 9-61).

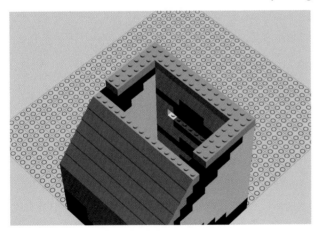

Figure 9-60. *Step 55: Adding seven 2x4 bricks*

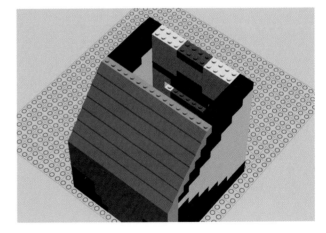

Figure 9-61. *Step 56: Adding yet more slopes*

57. Add two 2x3, two 2x4, and one 2x2 translucent slopes (Figure 9-62).

58. Add six 2x4 translucent bricks (Figure 9-63).

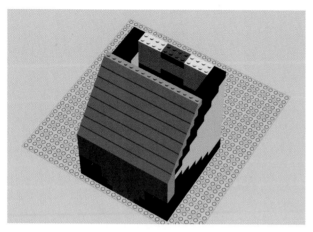

Figure 9-62. *Step 57: Adding another row of slopes*

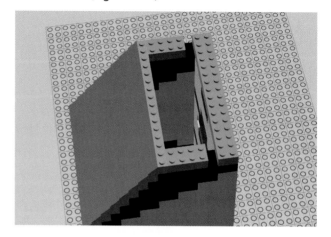

Figure 9-63. *Step 58: Attaching six 2x4 translucent bricks*

59. Next, you need two 2x1 translucent bricks (Figure 9-64).

60. Add six 2x2 and one 2x4 translucent slopes (Figure 9-65).

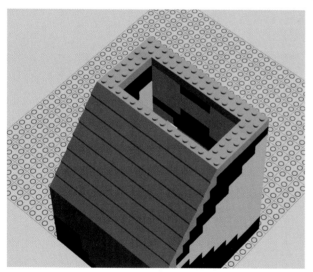

Figure 9-64. *Step 59: Adding two 2x1 translucent bricks*

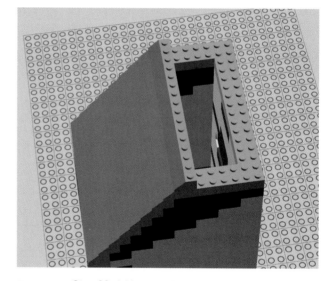

Figure 9-65. *Step 60: Adding another row of slopes*

Chapter 9

61. Add two 2x2 translucent bricks (Figure 9-66).

62. Now you need five 2x4 translucent bricks (Figure 9-67).

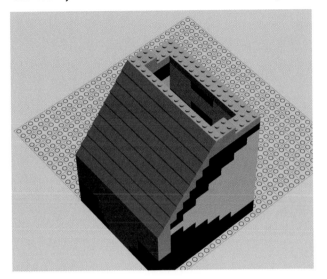

Figure 9-66. *Step 61: Adding two 2x2 translucent bricks*

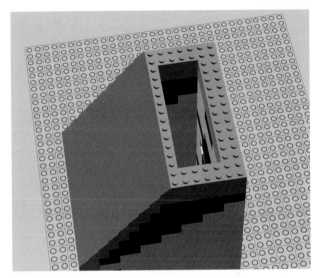

Figure 9-67. *Step 62: Adding five 2x4 translucent bricks*

63. Add five 2x2 and two 2x3 translucent slopes (Figure 9-68).

64. Add four 2x4 translucent bricks (Figure 9-69).

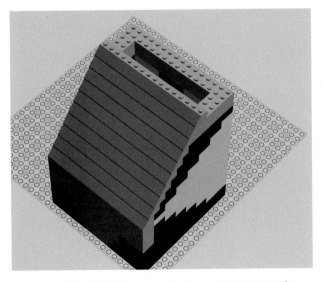

Figure 9-68. *Step 63: Will we never be done with this incline?*

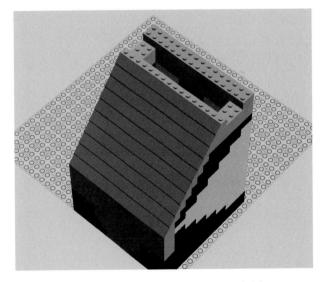

Figure 9-69. *Step 64: Adding four 2x4 translucent bricks*

65. Next, add two 2x1 translucent bricks (Figure 9-70).

66. Add two 2x2 translucent bricks (Figure 9-71).

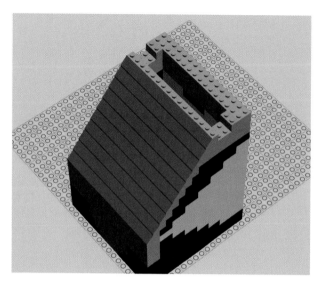

Figure 9-70. *Step 65: Adding two 2x1 translucent bricks*

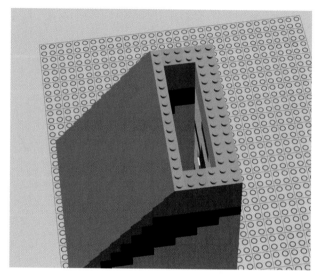

Figure 9-71. *Step 66: Completing the layer with two 2x2 translucent bricks*

67. Add six 2x2 and one 2x4 translucent slopes (Figure 9-72).

68. Add five 2x4 translucent slopes (Figure 9-73).

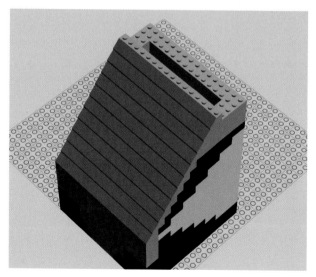

Figure 9-72. *Step 67: More slopes*

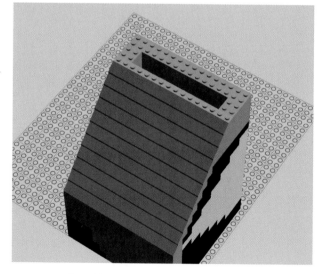

Figure 9-73. *Step 68: Adding five 2x4 translucent bricks*

69. Next, you need two 2x2 translucent bricks (Figure 9-74).

70. Add five 2x2 and two 2x3 translucent slopes. Almost done with slopes (Figure 9-75)!

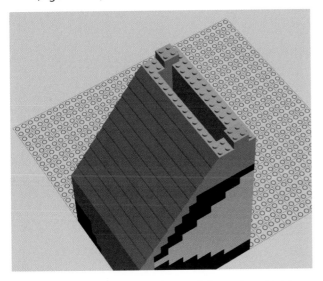

Figure 9-74. *Step 69: Attaching a pair of 2x2 translucent bricks*

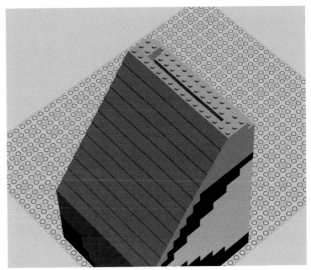

Figure 9-75. *Step 70: Placing some more slopes*

71. Next, add two 2x2 and two 2x4 translucent bricks (Figure 9-76).

72. Attach two 2x1 translucent bricks (Figure 9-77).

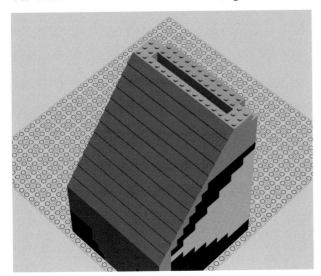

Figure 9-76. *Step 71: Attaching more bricks*

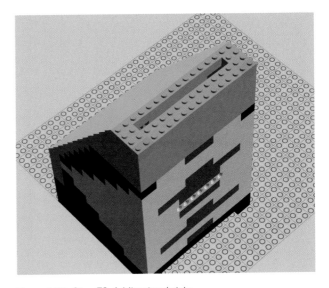

Figure 9-77. *Step 72: Adding two bricks*

73. Top 'er off with eight 2x4 translucent bricks (Figure 9-78).

74. Last slopes, we promise! We used 14 2x2 and one 2x4 transparent slopes (Figure 9-79).

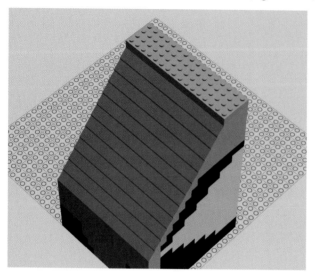

Figure 9-78. *Step 73: Adding eight 2x4 translucent bricks*

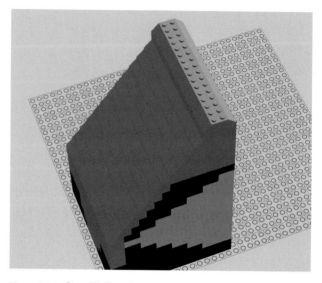

Figure 9-79. *Step 74: Translucent slopes add a decorative topper to the Lamp*

75. Reinforce the underside of the slopes with eleven 1x8 and one 1x4 translucent bricks (Figure 9-80).

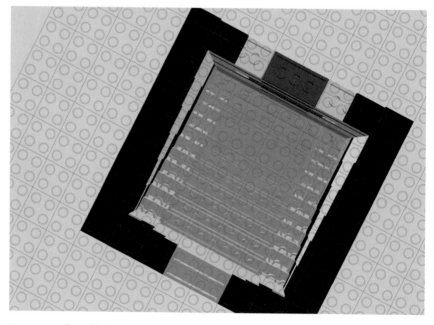

Figure 9-80. *Step 75: Reinforcing the underside of the slopes with these bricks*

76. Let's move on to the guts of the lamp. Grab two 3x5 angle beams and add seven connector pegs and a 3M peg as seen in Figure 9-81.

77. Set two more 3x5 angle beams next to the others. We'll secure them later! Also, add an 11M beam (Figure 9-82).

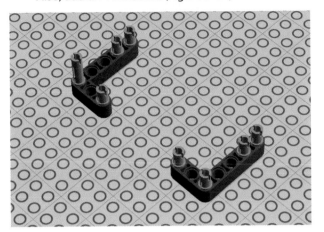

Figure 9-81. *Step 76: Inserting pegs into angle beams*

Figure 9-82. *Step 77: Adding a couple more angle beams and an 11M beam*

78. Add eight connector pegs to the beams (Figure 9-83).

79. Connect two more 11M beams as seen in Figure 9-84.

Figure 9-83. *Step 78: Inserting more pegs*

Figure 9-84. *Step 79: Adding 11M beams*

80. Add two 5M beams (Figure 9-85).

81. Add three connector pegs and two cross connectors (Figure 9-86).

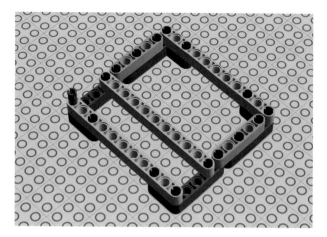

Figure 9-85. *Step 80: Adding 5M beams*

Figure 9-86. *Step 81: Inserting pegs and cross connectors*

82. Add a 9M beam as seen in Figure 9-87.

83. Attach a pair of 3M beams with pegs (Figure 9-88).

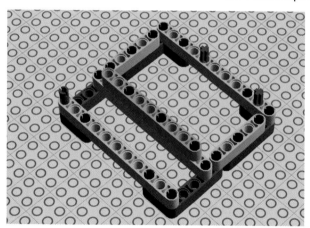

Figure 9-87. *Step 82: Adding a 9M beam*

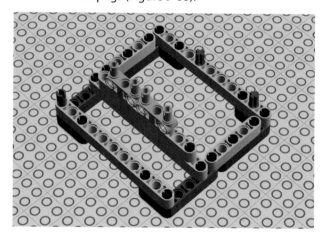

Figure 9-88. *Step 83: Attaching two 3M beams with pegs*

84. Add two 3M pegs to a motor (Figure 9-89).

85. Attach a 7M beam to the 3M pegs on the motor (Figure 9-90).

Figure 9-89. *Step 84: Adding a pair of 3M pegs to a motor*

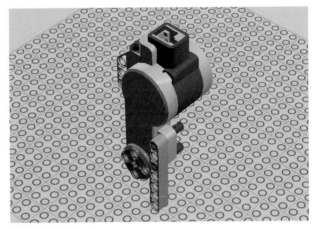

Figure 9-90. *Step 85: Adding a 7M beam*

86. Add two more 3M pegs as seen in Figure 9-91.

87. Add a 3x2 cross block (Figure 9-92).

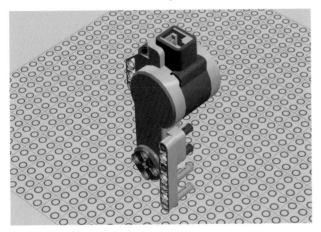

Figure 9-91. *Step 86: Inserting the pegs*

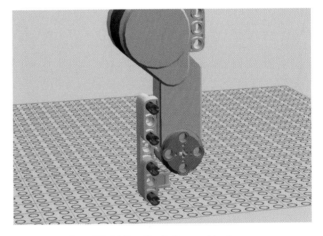

Figure 9-92. *Step 87: Adding the 3x2 cross block*

88. Secure the ends of the 3M pegs with another 7M beam (Figure 9-93).

89. Connect the motor assembly to one of the cross connectors (Figure 9-94).

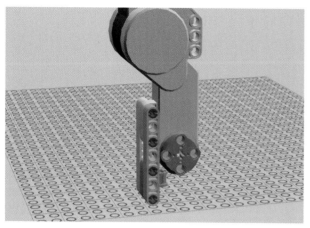

Figure 9-93. *Step 88: Adding a 7M beam*

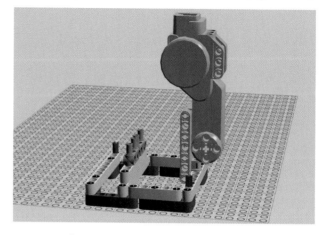

Figure 9-94. *Step 89: Connecting the motor assembly*

90. On the other cross connector, add a 0-degree angle element (Figure 9-95).

91. Add a cross connector to the angle element (Figure 9-96).

Figure 9-95. *Step 90: Adding the angle element*

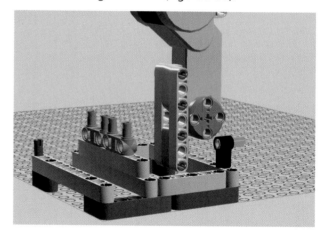

Figure 9-96. *Step 91: Adding a cross connector*

92. Add a cross block (Figure 9-97).

93. Connect a touch sensor as seen in Figure 9-98.

Figure 9-97. *Step 92: Adding a cross block*

Figure 9-98. *Step 93: Connecting the touch sensor*

94. Insert a pair of connector pegs as shown in Figure 9-99.

95. Add a 5M beam (Figure 9-100).

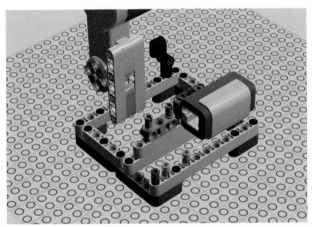

Figure 9-99. *Step 94: Adding pegs*

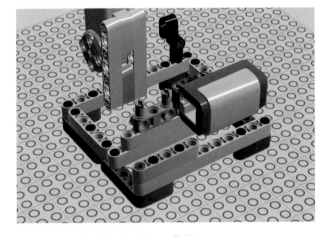

Figure 9-100. *Step 95: Attaching a 5M beam*

96. Attach another 3M beam with pegs (Figure 9-101).

97. Attach four tubes as shown in Figure 9-102. We're almost done!

Figure 9-101. *Step 96: Adding a 3M beam with pegs*

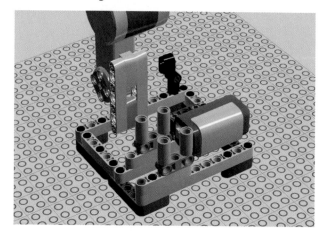

Figure 9-102. *Step 97: Adding tubes*

98. Insert four cross connectors into the tubes (Figure 9-103).

99. Add 0-degree angle elements (Figure 9-104).

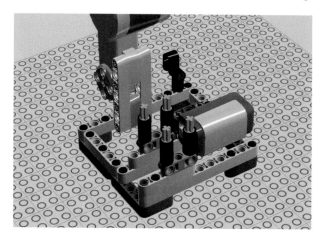

Figure 9-103. *Step 98: Inserting cross connectors*

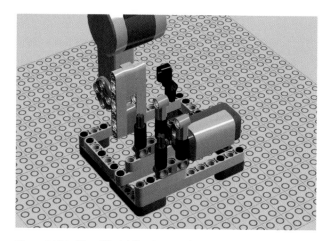

Figure 9-104. *Step 99: Adding two angle elements*

100. Add the dampers (the little rubber parts) to the free cross connectors, as seen in Figure 9-105. These will help nestle the night light. This is the time when you should add the night light and secure it with zip ties. To see how this should look, take a peek at Figure 9-111.

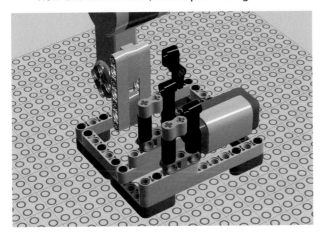

Figure 9-105. *Step 100: Adding dampers*

101. Put the guts of the clock inside and thread the cross axles through the Technic holes in the front as shown in Figure 9-106. The shorter axle should fit into the business end of the touch sensor, while the longer one should slide through the 0-degree angle element and into the motor's hub.

102. Add a ball element to the touch sensor's axle and a 4-tooth gear to the motor's axle (Figure 9-107).

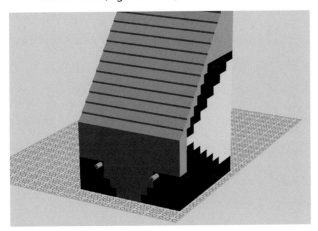

Figure 9-106. *Step 101: Putting the guts inside and threading cross axles*

Figure 9-107. *Step 102: Adding ball element and gear*

103. OMG! You're done! Your lamp should look like the rendering in Figure 9-108.

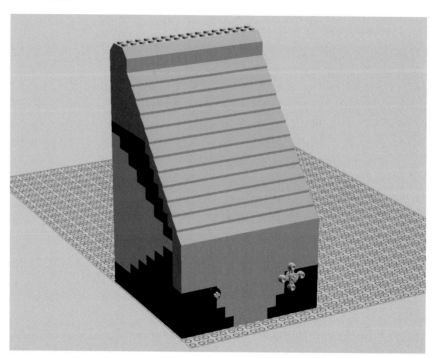

Figure 9-108. *Step 103: You're done!*

Install the Electronics

To wire up the Lamp, follow these steps:

- Attach the Arduino to its mounting plate and seat the Seeed Bluetooth Shield to the Arduino with the Bricktronics Shield on top of it. So, the Seeed shield will be sandwiched between the Bricktronics Shield and the Uno.

- Attach the mounting plate to the back of the Lamp and secure with half bushes.

- Seat the night light on the Technic assembly (Figure 9-111), using the rubber parts to keep it steady. Secure with a zip tie. (You'll need to remove the assembly from the inside of the lamp first, by pulling out the cross axles.)

- Connect the motor and touch sensor to the shield via Mindstorms cables.

- Plug in the PowerSwitch Tails (Figure 9-110) to one another and to the nightlight as seen in Figure 9-109. Don't forget the resistor!

- Wire up the tails and secure the wires to the cord with your Vaisis spiral wiring harness.

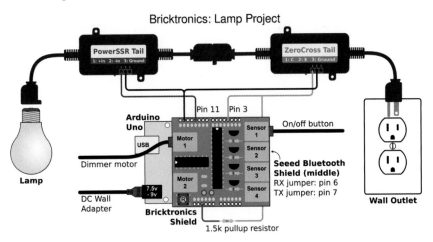

Figure 9-109. *Wire up the lamp as you see here*

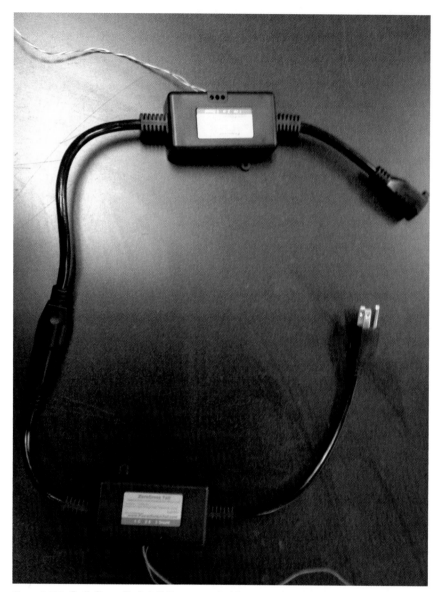

Figure 9-110. *Both PowerSwitch Tails are needed if you want to dim the lamp*

Chapter 9

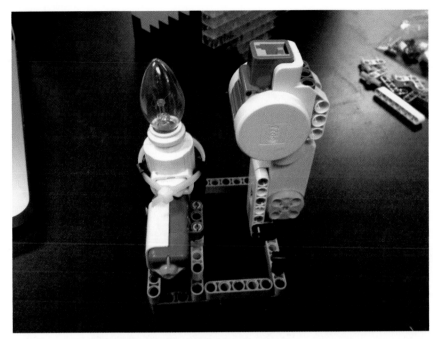

Figure 9-111. *The night light offers illumination in a small size*

Download and Install the App

We've written an Android app (shown in Figure 9-112) that works with most Android phones and tablets. There have been some reports that the Bluetooth Shield we use in this project doesn't work with a few Android phones—check the Bricktronics page (*http://wayneandlayne.com/bricktronics*) for compatibility information. The app is available on the Google Play store, as well as on the Bricktronics website.

If your device doesn't support the Google Play store, you'll have to enable "Install apps from unknown sources" in your settings. It is usually located in the "Applications" category. Download the app from the Bricktronics website. Open the *.apk* file—this can be done in a variety of ways, but clicking on the "download completed" notification should work on most devices.

Once you have it installed, go into your Android device's settings, turn on Bluetooth, and search for Bluetooth devices. If the lamp is plugged in and the Arduino is programmed, your Android device should find the lamp. Pair with the lamp (the default pairing PIN code should be 1234; if not, try 0000, 12345, or check the documentation at the site where you bought the Bluetooth module).

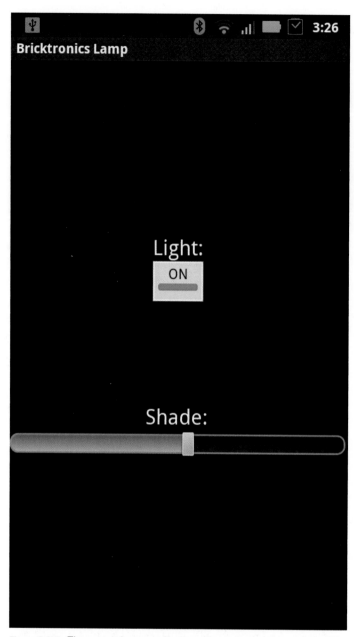

Figure 9-112. *The smartphone application allows you to turn on and off the Lamp from your phone, as well as dim it*

Once you've done this setup once, you won't have to do it again unless you switch Android devices. To use the Bricktronics Lamp app, open it, press the menu button, and then choose "connect with a device." After you've connected with the lamp, you can press the light button or slide the dimness slider. It will toggle the light or set the dimness—and the knob will even turn on the night light!

If you don't have an Android device, you can still activate the lamp over Bluetooth—you just won't have a UI. If you have Bluetooth on your computer, you can pair the lamp with your computer. The exact steps vary depending on your operating system. After it is successfully paired, your computer should have an additional serial connection or COM port. If you select that COM port in the Arduino software, you can use the Serial Console to control the lamp. In the Serial Console, if you type "l" in the text input, and then press Send, the light should turn on. If you type "m" in the text input, and then press Send, the light should turn off. (Make sure to have the Serial Console set to "no line ending.") To set the dimness, you'll have to consult an ASCII table, which shows the ASCII conversion of character to numeric values. Any value between 0 and 99 will set the dimness, with 0 being brightest and 99 being dimmest. If you look at an ASCII table, that means "!" should be pretty bright, and "d" should be practically off. There's a bunch of punctuation, numbers, and capital letters in between.

Program the Lamp

The Lamp sketch is surprisingly complicated. It must monitor a serial port and watch a zero-cross output, and make sure it starts the TRIAC at the right time. To do all this at the same time and keep the code readable, we're using *interrupts*. An interrupt is exactly what its name suggests—a way for the currently running program to be interrupted. These sound esoteric, and they really do help code work well and stay readable, but they can introduce errors if you're not used to thinking of your code being interrupted.

We attach the zero-cross line to digital pin 3, and attach an interrupt to it using `attachInterrupt()`. Every time this line goes from 5V to 0V, or "falls," we'll run a function. Similarly, we're going to attach a function to the Timer 1 interrupt using the TimerOne library. This is an easy way to make sure that a function is called at a specific period, and it's a bit more accurate than the

```
while (millis() < end_time) {}
```

idiom we've used throughout the book.

We define a simple protocol between the Arduino and the remote Bluetooth device. This is a simple, one-byte protocol. Sending the ASCII character q is a query, asking for the status of the remote device. The ASCII character l indicates lit, and the ASCII character m indicates off. Sent to the Arduino, they indicate commands, and sent from the Arduino, they indicate status. Outside of this, sending the values 0 through 99 as a binary value indicates a dimness, with 0 indicating fully on, and 99 indicating fully off.

Once you've got it all set up, you press the button on the lamp, and it toggles the light fully on or fully off. If you turn the knob, it dims the light. If you have an Android device with the Bricktronics Lamp app, after it's been paired with the Bluetooth Shield, start the app, and connect (the app only connects to paired devices). Select the Bricktronics Lamp entry. Now you can press the light button or slide the dimness slider. It will toggle the light or set the dimness—and the knob will even turn on the night light! (If you don't have it connected through the Bluetooth Shield, everything will still work.)

If you don't have an Android device, don't despair! This project can be done without the Bluetooth Shield, which will also save a little money! If you're not interested in dimming, the project can be modified to use a regular PowerSwitch Tail, which will provide on-off control, but not dimming. This will also save money, but will require circuit and code modifications that we don't go into.

Upload the following code to your Uno:

```
#include <Wire.h>
#include <Adafruit_MCP23017.h>
#include <Bricktronics.h>
#include <SoftwareSerial.h>
#include <TimerOne.h>  ❶

// Make: Lego and Arduino Projects
// Chapter 9: Lamp
// Website: http://www.wayneandlayne.com/bricktronics/

//Based on BluetoothShield Demo Code PDE by Steve Chang
//at Seeed Technology, Inc. and
//AC Light Control by Ryan McLaughlin.

Bricktronics brick = Bricktronics();
PIDMotor m = PIDMotor(&brick, 1);  ❷
Button lamp_switch = Button(&brick, 1);  ❸

#define RxD 6  ❹
#define TxD 7  ❺
#define AC_pin 9  ❻

volatile int i = 0;
volatile boolean zero_cross = false;

int dim = 0;
int freqStep = 65;  ❼

long last_encoder_reading = 0;
boolean lamp_lit = false;
long last_released = 0;
int destination_updates;

SoftwareSerial bluetoothSerial(RxD,TxD);  ❽

void setup()  ❾
{
  Serial.begin(9600);
  setupBluetoothConnection();
  brick.begin();
  m.begin();
  lamp_switch.begin();

  m.mKP = 1;  ❿
  m.mKD = 1;
  m.mKI = 0.003;

  pinMode(AC_pin, OUTPUT);
  attachInterrupt(1, zero_cross_detect, FALLING);  ⓫
  Timer1.initialize(freqStep);  ⓬
  Timer1.attachInterrupt(dim_check, freqStep);  ⓭
```

```
}

void loop() ⑭
{
  char recvChar;
  if (bluetoothSerial.available()) ⑮
  {
    recvChar = bluetoothSerial.read();
    switch (recvChar)
    {
    case 'q': ⑯
      if (lamp_lit)
      {
        bluetoothSerial.print('l');
      } else
      {
        bluetoothSerial.print('m');
      }
      bluetoothSerial.print((char) dim);
      break;
    case 'm': ⑰
      lamp_lit = false;
      dim = 128;
      break;
    case 'l': ⑱
      lamp_lit = true;
      dim = 0;
      break;
    default: ⑲
      if (recvChar >= 0 && recvChar <=99) ⑳
      {
        Serial.println("Setting dimness:");
        dim = recvChar;
        destination_updates = 20; ㉑
        m.go_to_pos(dim*4); ㉒
      }
    }
  }

  if (destination_updates) ㉓
  {
      m.update();
      delay(25);
      destination_updates -= 1;
  } else
  {
    m.stop(); ㉔
  }

  long current_encoder_reading = m.encoder->read();

  if (current_encoder_reading != last_encoder_reading) ㉕
  {
    dim = current_encoder_reading / 4; ㉖
    Serial.print("dim: "); Serial.println(dim); ㉗
    dim = constrain(dim, 0, 100);
    bluetoothSerial.write((byte)dim); ㉙
    last_encoder_reading = current_encoder_reading;
  }
```

```
            if (! lamp_switch.is_pressed())
            {
              last_released = millis();
            }
            else if (lamp_switch.is_pressed() && last_released > 0
                    && last_released - millis() > 50) ㉙
            {
              Serial.println("pressed"); ㉚
              last_released = 0;
              if (lamp_lit)
              {
                lamp_lit = false;
                dim = 128;
                bluetoothSerial.print("m"); ㉚
              }
              else {
                lamp_lit = true;
                dim = 0;
                bluetoothSerial.print("l"); ㉛
              }
            }

            if (Serial.available()) ㉜
            {
              recvChar  = Serial.read();
              bluetoothSerial.print(recvChar);
            }
          }

          void setupBluetoothConnection()
          {
            bluetoothSerial.begin(9600); ㉞
            bluetoothSerial.print("\r\n+STWMOD=0\r\n"); ㉟
            bluetoothSerial.print("\r\n+STNA=Bricktronics\r\n"); ㊱
            bluetoothSerial.print("\r\n+STOAUT=1\r\n"); ㊲
            bluetoothSerial.print("\r\n+STAUTO=0\r\n"); ㊳
            delay(2000); ㊴
            bluetoothSerial.print("\r\n+INQ=1\r\n"); ㊵
            Serial.println("The Bluetooth module should be connectable!");
            delay(2000); ㊶
            bluetoothSerial.flush(); ㊷
          }

          void zero_cross_detect() ㊸
          {
            zero_cross = true;
          }

          void dim_check()
          {
            if (zero_cross && lamp_lit) ㊹
            {
              if (i >= dim) ㊺
              {
                digitalWrite(AC_pin, HIGH); ㊻
                delayMicroseconds(5); ㊼
                digitalWrite(AC_pin, LOW); ㊽
                i = 0; ㊾
                zero_cross = false;
```

```
      }
   else
      {
        i++;  ㊿
      }
    }
 }
```

❶ TimerOne is an Arduino library, and can be downloaded at *http://www.arduino.cc/playground/Code/Timer1*.

❷ Plug a motor into Motor Port 1.

❸ Plug a button into Sensor Port 1.

❹ Connect the jumpers on the Bluetooth Shield so RX is connected to pin 6.

❺ Connect the jumpers on the Bluetooth Shield so TX is connected to pin 7.

❻ Connect the PowerSSR Tail input pin to pin 9 on the Arduino. While you're at it, make sure you've connected the Zero Cross output pin to pin 3 on the Arduino.

❼ `freqStep` is the delay between dimness checks. 65 works well for 60 Hz power, and 78 works well for 50 Hz power.

❽ This creates a "software driven serial port" on pins RxD and TxD. It can't go as fast as the built-in Serial port, but it works well for what we need.

❾ `setup()` runs once at startup.

❿ This changes the KP, KD, and KI tuning numbers on the PID for the motor object m.

⓫ Every time the signal on pin 3 goes from high to low, run the function `zero_cross_detect`.

⓬ Initialize the Timer1 library with the period `freqStep`.

⓭ Every `freqStep` milliseconds, run the function `dim_check`.

⓮ `loop()` runs over and over.

⓯ Check to see if we have received a character from the Bluetooth module.

⓰ If the character was "q," it's a query from the Bluetooth module, asking us what our current status is.

⓱ If the character was "m," the light should be turned off completely.

⓲ If the character was "l," the light should be turned on fully.

⓳ If the received character hasn't been handled by now, it's either garbage, or a dimness setting.

⓴ If the character has a value between 0 and 99, treat it as a dimness value.

㉑ The next 20 times `loop()` is run, try to get the motor in the dimmer knob close to the destination.

㉒ Set the motor's destination to the dimness value sent times 0 to 99 on the knob is not very far—0 to 396 is much better.

㉓ If `destination_updates` is greater than zero, then try to get the motor closer to the destination for a little bit, before managing the rest of the logic.

㉔ Make sure the motor stops after it's tried to get to its destination.

㉕ If the knob has turned since the last run through `loop()`, then everything in the next three steps happens.

㉖ This compensates for the increase in range when having the knob remotely set.

㉗ Update the Serial console to the PC with some debug information.

㉘ Update the remote Bluetooth device with the new dimness value.

㉙ If the lamp switch is pressed, and it isn't the first time it was pressed, and the last time it was let go was at least 50 milliseconds ago, this means we want to toggle the light status!

㉚ Update the Arduino Serial Monitor with debug information.

㉛ Update the remote side with the new status that the light is off.

㉜ Update the remote side with the new status that the light is on.

㉝ If the user sends data to the Arduino Serial Monitor, it's a debugging command from the user, and we want to forward that onto the remote Bluetooth device. This lets you control the Bluetooth software through the Arduino Serial Monitor.

㉞ This is the start of the function that configures the Bluetooth module. Set the software serial link to 9600 baud.

㉟ Tell the Bluetooth module to switch to slave mode.

㊱ Tell the Bluetooth module to change its name to Bricktronics.

㊲ Tell the Bluetooth module to permit paired connections.

㊳ Tell the Bluetooth module to deny automatic connections.

㊴ Give the Bluetooth module time to change its settings.

㊵ Tell the Bluetooth module to enable connection requests.

㊶ Give the Bluetooth module time to change its settings.

㊷ Make sure all the data is sent, just in case something got stuck in the Software Serial output queue. This is the end of the function that configures the Bluetooth module.

㊸ `zero_cross_detect` is called when a zero cross is detected, and it sets a global flag, `zero_cross`, to true.

㊹ If the lamp is supposed to be on, and there's been a zero-crossing since the last one we handled, then we go on to the next step.

㊺ If we've gotten as far or further into the waveform as we need to get the dimness we want, then we do everything in the next four steps.

46 Turn on the PowerSSR Tail TRIAC.

47 Wait to make sure it turns on.

48 Stop turning on the TRIAC. It will turn off on its own at the next zero cross.

49 Reset `i`, and the `zero_cross` boolean.

50 It isn't time to turn off yet, so keep track of how many times we've checked.

The Next Chapter

We're done with models! In Chapter Ten we'll cover certain Advanced Techniques involving breadboarding up Bricktronics substitutes, learning about XBee wireless modules, discovering methods of powering your robots, and we'll even cover motors in depth!

Advanced Techniques 10

In this chapter we'll explore a number of advanced topics we couldn't fit into the other parts of the book. We'll show you how to create your own Bricktronics-compatible boards if you don't want to buy them, we'll describe all the intricacies of motors, we'll go in-depth on the ways Arduinos and NXT bricks can communicate, and as if that weren't enough, we'll explore the technology of XBee wireless boards.

Wiring Up Bricktronics Equivalents

We've assumed you're using the Bricktronics Shield and the Bricktronics Motor Controller throughout the book, but you don't have to. Not only can you recreate them on breadboards, the Bricktronics Motor Controller and the Bricktronics Shield are open source hardware created by Wayne and Layne, LLC (co-authors Adam and Matthew). That means that we have released the files needed to create the hardware, and legally relaxed the rights so you can use the files. Depending upon the exact license used, the rights can vary. These are licensed under a Creative Commons Attribution-Share Alike 3.0 License. Legally, this means you may copy, distribute, and display them (and any modifications or "derivative works" you make) if you give credit to Wayne and Layne and you must also license any modifications or derivative works under this exact same Creative Commons Attribution-Share Alike 3.0 License. This means you can go to the Bricktronics website, download our PCB files (Figure 10-1), and do what you'd like with them. You can send them to a manufacturer to have boards made, or you can try to mill or etch your own. You can download the files, mash them up with other boards, post them all over the Internet—as long as you preserve the license.

Knowing that not everyone wants to use PCBs, we designed the hardware to be as breadboard-able as possible. After all, many of the projects in this book were first tested with a small nest of wires connecting a few chips on a breadboard. To recreate a Bricktronics Shield on a breadboard (Figure 10-2), you'll need an L293D H-bridge chip, an MCP23017 I2C I/O port expander, some capacitors, and some wires. You'll still need an Arduino to plug into. If you're recreating the Chocolate Milk Maker, you'll need two TIP120 transistors.

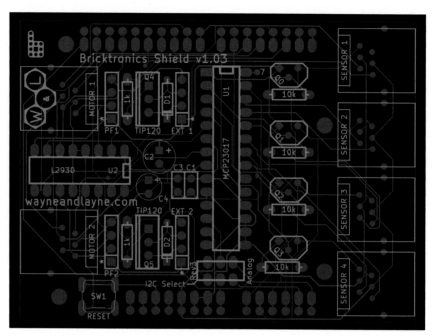

Figure 10-1. *Want to etch your own Bricktronics boards? Go right ahead!*

The Molex pins we used were P/N 08-50-0113, and they work with the KK series, which has a full line of terminal housings, including a 4 pin housing, Molex P/N 22-01-3047, and a 6 pin housing, Molex P/N 09-50-8063. They are available at Mouser and Digi-Key, and many other suppliers. The double length male breakaway headers we used were Sullins PEC36SFCN, available at Digi-Key and other suppliers.

We've created a system diagram showing the various components and their connections. The connections for the sensors and motors have been included in a table, to keep the diagram readable. After breadboarding the L293D H-bridge chip and the MCP23017 I/O port expander, you'll have to connect your motors and sensors. Consult Table 10-1 to see how to connect your breadboard to Lego components.

Table 10-1. Motor and sensor connections

NXT Motor 1	White, Control A, pin 1	L293D pin 3
	Black, Control B, pin 2	L293D pin 6
	Red, Ground, pin 3	Ground
	Green, Power, pin 4	5 V
	Yellow, Encoder A, pin 5	Arduino pin 2
	Blue, Encoder B, pin 6	Arduino pin 5
NXT Motor 2	White, Control A, pin 1	L293D pin 11
	Black, Control B, pin 2	L293D pin 14
	Red, Ground, pin 3	Ground
	Green, Power, pin 4	5 V
	Yellow, Encoder A, pin 5	Arduino pin 3
	Blue, Encoder B, pin 6	Arduino pin 4

PF Motor 1	9V, pin 4	9V
	Control B, pin 3	L293D pin 3
	Control A, pin 2	L293D pin 6
	Ground, pin 1	Ground
PF Motor 2	9V, pin 4	9V
	Control B, pin 3	L293D pin 11
	Control A, pin 2	L293D pin 14
	Ground, pin 1	Ground
NXT Sensor 1	White, ANALOG, pin 1	Arduino pin A0
	Black, GROUND, pin 2	Ground
	Red, GROUND, pin 3	Ground
	Green, IPOWERA, pin 4	5 V
	Yellow, DIGIAI0, pin 5	Arduino pin 8
	Blue, DIGIAI1, pin 6	Arduino pin 12
NXT Sensor 2	White, ANALOG, pin 1	Arduino pin A1
	Black, GROUND, pin 2	Ground
	Red, GROUND, pin 3	Ground
	Green, IPOWERA, pin 4	5 V
	Yellow, DIGIAI0, pin 5	IO Expander pin 26
	Blue, DIGIAI1, pin 6	IO Expander pin 25
NXT Sensor 3	White, ANALOG, pin 1	Arduino pin A2
	Black, GROUND, pin 2	Ground
	Red, GROUND, pin 3	Ground
	Green, IPOWERA, pin 4	5 V
	Yellow, DIGIAI0, pin 5	IO Expander pin 24
	Blue, DIGIAI1, pin 6	IO Expander pin 23
NXT Sensor 4	White, ANALOG, pin 1	Arduino pin A3
	Black, GROUND, pin 2	Ground
	Red, GROUND, pin 3	Ground
	Green, IPOWERA, pin 4	5 V
	Yellow, DIGIAI0, pin 5	IO Expander pin 22
	Blue, DIGIAI1, pin 6	IO Expander pin 21

It's important to be methodical and careful when breadboarding projects in this book. You shouldn't try to change connections while the breadboard is powered up—you're just asking to fry components! That being said, we don't want to scare you off—we breadboarded these projects (Figure 10-3) while we were developing them.

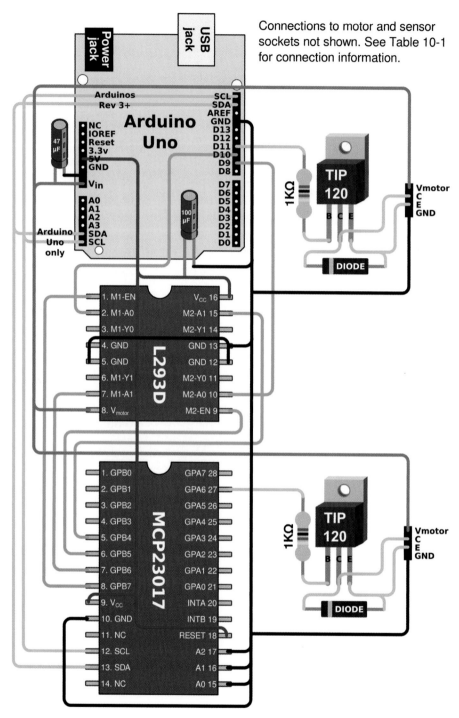

Connections to motor and sensor sockets not shown. See Table 10-1 for connection information.

Figure 10-2. *Breadboarding a Bricktronics shield is easy*

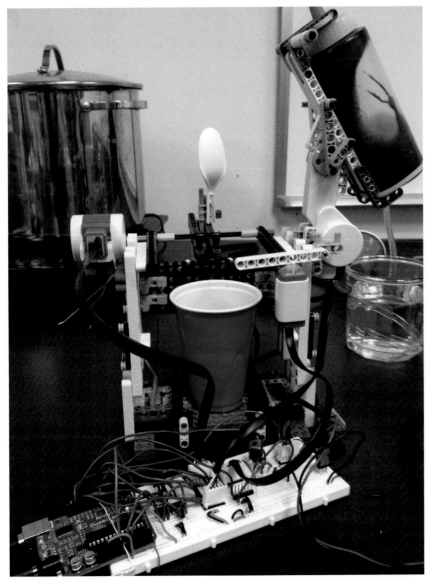

Figure 10-3. *If you don't want to buy Bricktronics boards, you can always breadboard up the equivalent—we did!*

Mounting PCBs to Legos

Let's be realistic: it's kind of hard to connect non-Lego parts to Lego parts. The solution is to create mounting plates (Figure 10-4) out of wood or acrylic. This is how it works:

1. Download the plate patterns from *http://wayneandlayne.com/bricktronics* and cut them out using a laser cutter—or if you don't have access to one, you can use a saw and drill to make a similar plate. Alternatively, we sell the plates! Go to Wayne and Layne's website or the Maker Shed and search for Bricktronics.

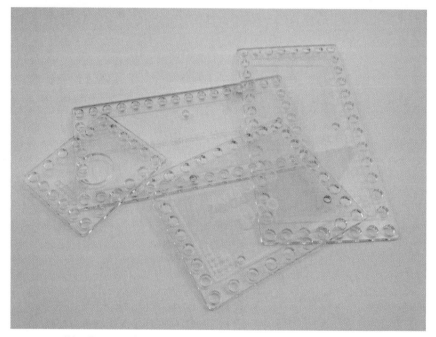

Figure 10-4. *We offer a number of mounting plates to help you combine Legos and Arduinos!*

2. Next, we need to modify screws (Figure 10-5) to attach the PCB to the plate. Grab some screws (we used #6, 1/2" wood screws) and grind off the points as shown in Figure 10-5—you won't need them and they'll just poke through the plate and maybe scratch your hand or a model.

3. The plates' holes are spaced like Technic beams, so you can connect the plates directly to your robot! First, find a good area to mount the plate, with plenty of free holes. (We added mounting surfaces to all of the robots we built for this book.) Next, slide some cross connectors into the appropriate holes and then attach the plate as seen in Figure 10-6. Secure it with some half bushes and you're golden!

Figure 10-5. *Modify 1/2" screws to mount the Arduino to the plate*

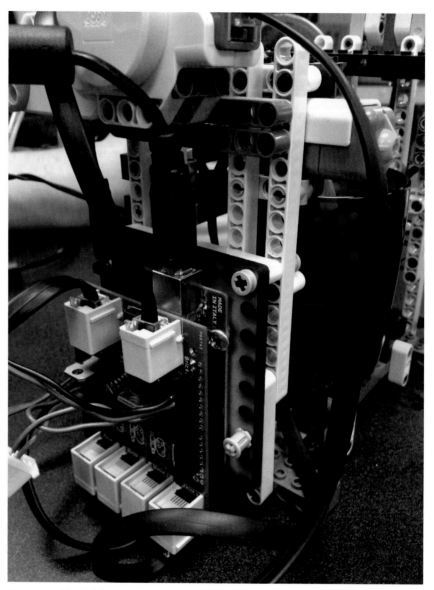

Figure 10-8. *The Bricktronics shield mounts onto a robot thanks to laser-cut mounting plates*

Adding Molex Connectors to Lego Wires

Various lengths of Mindstorms cable are included in the set and are used for the transfer of both electricity and data. In this section we'll learn how to add a Molex connector to allow us to plug the wires into a breadboard.

First, grab a 50 cm cable and cut it in half, giving you two 25 cm cables, each with a Mindstorms plug on one end of the wire. Next use a wire stripper to carefully nick the black coating around the inner wires on the cut side. Pull the coating off, checking to make sure you didn't cut the wires inside. If you did, pull the black coating back a little further. Strip the inner insulation off of each wire, and crimp a pin onto each wire as shown in Figure 10-7. Insert the wires in order into the terminal housing.

If you have a crimping tool for the pins, you can use that. If not, you can crimp the pins using a pair of pliers—it may not be reliable enough for the Space Shuttle, but it will do.

Similarly, you can add Molex connectors to Power Functions wires, as shown in Figure 10-8. Begin with an official Power Functions extension wire (P/N 8871) and cut it in half. Carefully cut and pull apart the wire so each of the four internal wires is exposed. Gently strip a bit of insulation off each wire, and crimp a pin onto each wire using a crimping tool.

Another option is to use Bricktronics breakout boards. We've produced a small PCB that has an NXT jack and 6 header pins. It can plug right into a breadboard, and eliminates the need to cut any NXT cables in half. They're available on our website.

The Molex pins we used were P/N 08-50-0113, and they work with the KK series, which has a full line of terminal housings, including a 4 pin housing, Molex P/N 22-01-3047, and a 6 pin housing, Molex P/N 09-50-8063. They are available at Mouser and Digi-Key, and many other suppliers.

Figure 10-7. *The standard Mindstorms cable consists of six wires*

Figure 10-8. *Power Functions wires transmit power and control signals*

All About Motors

Next, let's discuss motors. You'll find yourselves using motors all the time if you dabble in robotics, so you might as well learn all you can about them!

There are a variety of motors you may encounter working with hobbyist electronics. The three main motor types you'll find are *DC brushed motors*, *stepper motors*, and hobby *servo motors*. With the exception of many servo motors, you can't just hook a motor up to an Arduino. The other motors take more current than an Arduino can source from a pin.

DC Brushed Motors

DC brushed motors (Figure 10-9) are great for providing motion without precision. These motors come in all shapes and sizes. They have two terminals. When you put a voltage across the two terminals, the shaft spins. A higher voltage speeds it up and a lower voltage slows it down—with limits. Giving it more voltage than it's rated for can damage it, and there's a minimum voltage where it won't spin as well. Make sure to check the motor's datasheet to find these numbers. You can usually find a datasheet on the same website you bought the motor from.

To reverse the direction, all you have to do is reverse the polarity across the two terminals.

Figure 10-9. *DC brushed motors are the simplest and most common sort of motor*

Driving a DC brushed motor

DC brushed motors need more current than an Arduino pin can source. The two easiest ways to drive one of these motors are with a transistor or with an H-bridge. A single transistor can be connected to the power source for the motor, an Arduino output pin and a motor terminal. The other motor terminal is connected to ground. The output pin then controls the transistor, connecting and disconnecting the motor power source to the motor. When the motor is connected, it spins one direction. When it isn't, it stops. If the output pin is capable of PWM (Pulse Width Modulation, described in "About the L293D Chip" in Chapter Four), then the motor can be powered up and powered down fast enough that the motor keeps spinning, but at a slower rate.

In order to change direction, the motor terminals need to switch polarity: the ground terminal needs to be powered, and the powered terminal needs to go to ground. The easiest way to do this is with a circuit configuration called an *H-bridge*. This configuration has two inputs and two outputs. The two inputs can control the speed and direction of a DC brushed motor connected across the two outputs. H-bridges often come in integrated circuits, like the L293D (see "About the L293D Chip" in Chapter Four).

Flyback Diodes

A *voltage spike* occurs when a DC brushed motor (or any other inductive load) stops. This is caused by collapsing electromagnetic fields. This is called *flyback*, or sometimes called *back EMF* (electromotive force). The voltage spikes can cause damage, especially to relatively delicate components like microcontrollers. To prevent this, a diode is usually placed backwards across the motor terminals. This is called a *flyback diode*. When driving a DC brushed motor, make sure to account for the flyback. The L293D H-bridge chip has internal flyback diodes.

Stepper Motors

Stepper motors (Figure 10-10), also known as steppers, are motors that move step-by-step. Each step forward or backward advances the motor a specific angle. They are often used for precision control, for things like 3D printers and CNC machines.

Figure 10-10. *Steppers offer precision control that other motors lack*

A stepper motor has a few coils inside, and when the coils are charged in a certain order, the motor advances or retreats a step. The easiest way to control these is with a stepper driver chip. After connecting a few pins from the Arduino to the driver, usually one pin controls direction and another controls the number of steps to advance. The driver usually connects to another power supply to power the motor.

Because steppers often take a separate chip for control, people often use a motor controller shield (such as the Arduino Motor Shield, Maker Shed P/N MKSP12) instead of breadboarding a chip into a circuit.

Hobby Servo Motors

Servos (Figure 10-11) are generally small, geared motors with integrated feed-back, meaning that the motor has an encoder that constantly informs the microcontroller about the hub's angle. We discuss Mindstorms motors, which behave similarly, in Chapter Two.

Servos often have a limited range of motion—0 to 90 degrees and 0 to 180 degrees are common. They have a three-pin interface: ground, power, and input. The input line is PWM, and controls the angle of the servo shaft. At one extreme, when the input line is tied to ground, the motor turns all the way to one end and stops. At the other end, when the input line is tied to power, the motor turns all the way to the other end and stops. The angle can be adjusted in between the two extremes by using PWM. There is an excellent library for Arduino called Servo that controls up to twelve servos on most Arduinos, and up to 48 servos on an Arduino Mega!

Figure 10-11. *A typical hobby servo motor*

Hobby servos are also used in RC aircraft. If you'd like to pick up some hobby servo motors and don't want to order online, you can probably find them at a local hobby shop that has RC aircraft supplies. You shouldn't need to spend more than $15 for a basic servo for your projects.

Powering Your Robot

In this section we'll describe the various options for powering your project. These consist of the DC power jack, which supplies electricity to the Arduino via a wall wart. There are USB connectors, which deliver both power and data through a USB cable. The third method is the battery pack, which we often use in this book. Let's go over each method.

DC Power Jack

A good choice for powering a Bricktronics project is an AC adapter, a so-called "wall wart" that plugs into the wall and supplies electricity to the board (Figure 10-12).

But not every wall wart is compatible with the Arduino! To work with the Arduino, the plug should be 2.1mm and center positive. To power both the Arduino and motors, the adapter should be 7.5 to 9V, and provide at least 600 mA per motor. While it's recommended that the Arduino be run from a 7.5V to 12V AC adapter, the NXT motors aren't meant to be run at anything more than 9V.

The voltage directly off the DC jack is made available at the VIN pin. Arduinos also have a 5V regulator, and when the DC jack is used, the power from the DC jack goes into the regulator, and the 5V output is connected to the 5V pin.

If you're looking for an Arduino-compatible wall wart, we suggest buying the 9V power supply with the part number MKSF3 from the Maker Shed.

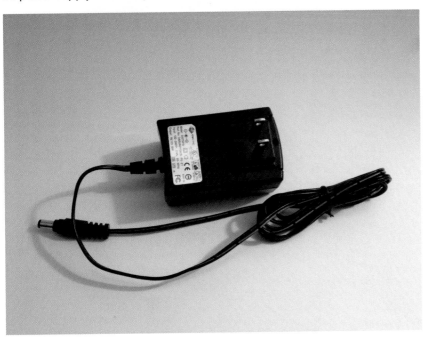

Figure 10-12. *The wall wart powers your Arduino, but doesn't help manage data*

USB Connector

A standard Arduino has a USB connector (Figure 10-13), through which it can power the board. It generally provides up to 500 mA at 5 V. This is more than enough for the Arduino and most peripherals, but NXT motors take around 600 mA at 7.2 V to 9 V, so an Arduino with an NXT motor will not be suitably powered by a USB connection.

It's important to only supply power from one source to a circuit. Even if two supplies are nominally the same voltage, like the 5 V from USB and the 5 V from a voltage regulator, they are almost certainly not exactly the same, so there will be a voltage difference between the two supplies and this can cause issues.

Since the Duemilanove, the Arduino will automatically switch between the USB power and the DC power jack. In older Arduinos, or alternative boards, there may be a switch or jumper to select between USB power and the DC jack. Most Arduinos also have a 3V3 pin. This pin supplies a small amount of current at 3.3 V. Depending on your specific board, the 3V3 pin can only source 80 mA or less.

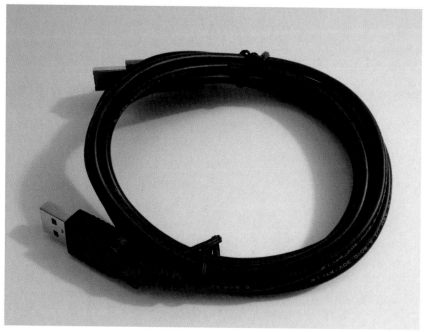

Figure 10-13. *USB is a great way to power your Arduino project—assuming you don't mind that it's tethered to the computer*

Batteries

Another option for powering your Arduino projects is by connecting a battery pack (Figure 10-14) to the DC jack. But how do you know which batteries to use?

Consumer batteries often have their capacity labeled in milliamp-hours or mAh. A cell with a capacity of 2300 mAh and a voltage of 1.2 volts can theoretically run a 1 milliamp load for about 2300 hours, or a 100 milliamp load for about 23 hours. When batteries are combined in series, their voltages add. Two cells like that, in series, could theoretically run that same 10 milliamp load at 2.4 volts for 230 hours.

In practice, however, there are many factors that reduce the capacity of a battery. The voltage of the battery decreases as its charge decreases. Depending on the battery type, there may be significant self-discharge. A battery is rated for a certain current draw, and drawing significantly more current than that comes at a cost. While there is a lot of information about battery technology available, a good way to estimate battery life is to try it—as long as you're not overcharging, shorting, or bypassing any safety measures on your batteries.

Normal *alkaline batteries*, like AAAs, AAs, and C- and D-cells, have a standard voltage of 1.5 V. When they're "dead," they normally have a voltage around 1 V. They're not usually rechargeable, but they do have a low self-discharge rate. This means that they can keep the same charge when they're not being used, even over months or years. The difference between the different alkaline battery sizes, like AA or C, is capacity. D cell batteries can generally power the same load for longer than a C cell can, all the way down to small AAAA cells.

The standard rechargeable battery in 2012 is a *nickel-metal hydride* (NiMH) battery. They have a significantly higher capacity than the older nickel-cadmium (NiCd) batteries. They have a standard voltage of around 1.25 V. When they're "dead," they normally have a voltage around 0.8 to 1 V. Most NiMH batteries have a relatively high self-discharge rate. A fully charged standard NiMH battery will not be fully charged a year later, even if it isn't plugged into anything! It can be recharged, of course, but this can be important to keep in mind. If you have a fire detecting robot, you probably wouldn't want to power it with standard NiMH batteries!

Lithium batteries are often used in low-power, long life situations. They are small, light, and have a low self-discharge. However, they have a high internal resistance, which means they cannot supply large amounts of current. Most lithium batteries have a voltage of around 3 V, and the common lithium CR2302 coin cell battery usually has a capacity of around 230 mAh.

Lithium-ion (LiIon) and *Lithium-polymer* (LiPoly or LiPo) are often the "non-replaceable" rechargeable batteries in consumer electronics. They are lightweight and can provide a lot of current. Lithium-ion batteries often come in hard cases. Lithium-polymer batteries are usually thin and rectangular. They are softer and easier to damage physically.

Battery Dos and Don'ts

There are a variety of voltages of cells. Make sure your charger voltage matches the voltage of the battery. Overcharging them will, at best, damage your battery, and at worst, start a fire!

Do not draw more current than the battery is rated to provide, and do not overcharge them or drain them below their minimum safe voltage. Many lithium rechargeable batteries have protection circuits attached to them that monitor those things for you, and help stop the batteries from being damaged or starting a fire. Verify they have a protection circuit—don't assume a battery comes with one!

Connecting lithium cells in parallel is not generally safe. Different batteries have slightly different characteristics, which can lead to one battery charging another, which isn't monitored like a battery charger is. Connecting lithium cells in series is also not generally good. Different cells will drain at different rates. There are a large variety of capacities and voltages available relatively cheaply online, making it easy to purchase the capacity and voltage that your project needs.

Figure 10-14. *Battery packs are critical for powering a mobile robot*

NXT to Arduino Communication

So far, we haven't talked about programming the NXT brick. The NXT brick (Figure 10-15) has an Atmel ARM7 processor, a graphic display, a small speaker, three motor ports, and four sensor ports. It's more powerful than a standard Arduino. One of the possibilities is connecting the NXT brick and the Arduino together. We'll discuss one way of doing this with I2C, a common way of connecting peripherals to microcontrollers.

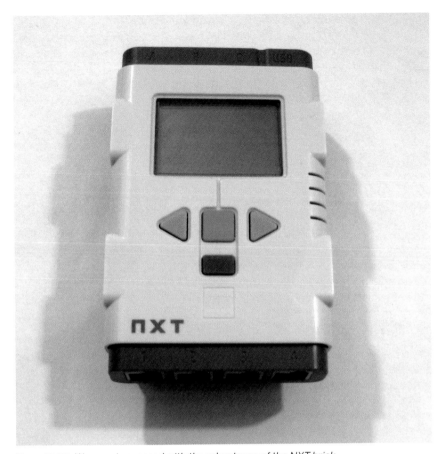

Figure 10-15. *We were impressed with the robustness of the NXT brick*

I2C is a two-wire interface for connecting to low-speed peripherals. It's fairly ubiquitous, and generally works well. It's used all over computing, from micro-controller boards to PC motherboards. It's a *bus protocol*, which means that a variety of devices can connect to the same two wires. Each device has an address on the bus. Generally, devices are either *masters* or *slaves* on the bus. Masters can read data from slaves, or write data to slaves, but slaves have no real in-protocol way of alerting masters of changes.

The Arduino has support for I2C in hardware, as well as being able to do I2C in software ("bitbanging"). The library for using the I2C hardware is called the Wire library. It supports making the Arduino a slave or a master. Many devices using I2C support the concept of a *register*, which is the data address at the slave side where a specific piece of data is stored.

The NXT brick uses I2C already—that's how the ultrasonic sensor communicates with the NXT. This means that an NXT and an Arduino can communicate.

However, default-configuration NXTs are programmed with a simplified language called NXT-G (Figure 10-16), a kid-friendly programming language that represents commands as "blocks." NXT-G programmers simply string together blocks to program their robots.

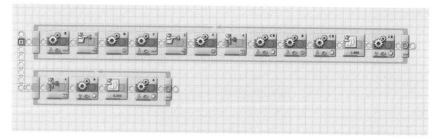

Figure 10-16. *NXT-G offers nonprogrammers a convenient, image-based programming environment—with limitations*

However, the downside of NXT-G is that it's impossible for nonprogrammers to create all-new blocks. As a result, some Mindstorms hackers have turned to alternative programming environments for the NXT.

We experimented with one of these, a language called NXC (*http://bricxcc. sourceforge.net/nbc/*). NXC or "Not eXactly C" is a C-like language for the NXT. It isn't quite as friendly as the language that Arduino uses, but there are a variety of examples that come with the compiler. There's an enhanced firmware for upgrading the NXT brick that provides multidimensional arrays and native shift operations to NXC, but our example here works fine without it.

To connect the two devices physically (Figure 10-17), we need to connect the SDA (I2C data) and SCL (I2C clock) lines together, as well as establish a common ground. On Arduino Unos Rev 2 and older, SDA is analog input 4, and SCL is analog input 5. On the Unos Rev 3 and newer, SDA and SCL are also located up near the AREF pin. These lines are planned to stay in that location across more of the Arduino line, simplifying I2C on Arduinos (for example, the Leonardo's analog 4 and 5 pins don't carry SDA and SCL, so you must use the new pins).

On the NXT, SDA is the blue wire, wire 6, and SCL is the yellow wire, wire 5. Cutting an NXT cable in half and crimping a connector on it works well. Then connect SDA on the Arduino to SDA on an NXT cable, and SCL on the Arduino to SCL on an NXT cable. Ground on the NXT wire is black, or pin 2; connect that to one of the Arduino ground pins.

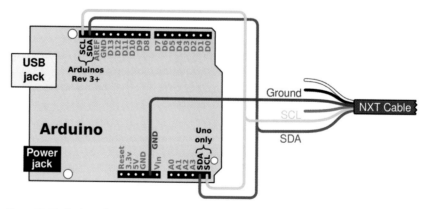

Figure 10-17. *System diagram*

Normally with I2C, there must be a *pullup resistor* on both SDA and SCL to allow the devices to communicate electrically. The Arduino has internal pullups. They are automatically turned on in the Wire library. Most regular I2C buses work best with stronger pullups than the Arduino internal pullups. The NXT has a different design, and actually wants slightly weaker pullups. In our testing, the internal Arduino pullups worked fine, but they might not work well if you make the I2C wires really long.

In NXC, we're keeping it as simple as possible. We set up I2C on port S1, wait for it to settle, and then we put the address of the Arduino and the register we want to interact with (which is arbitrary here, as the Arduino has no built-in requirements to use registers with I2C) in the byte array cmdbuf. Then we send that, plus any extra data bytes we want to send, to the Arduino. The NXT is expecting a byte in response, and that gets written from the Arduino into the outbuf array.

Here's the NXC code:

```
#define ID 0x01
#define ADDRESS ID
#define ADDRESS_SEND (ID << 1)
#define ADDRESS_RECV (ADDRESS_SEND+1)

task main(){
 byte addr = ADDRESS_SEND;
 byte reg = 0x43;
 byte num_of_bytes = 1;
 byte cmdbuf[];
 byte outbuf[];

 SetSensorType(S1, IN_TYPE_LOWSPEED);
 while (I2CCheckStatus(S1) == STAT_COMM_PENDING); // wait for S1 to settle
 Wait(100);

 ArrayBuild(cmdbuf, addr, reg);

 // if you want to send data to the Arduino, put it in cmdbuf
 // after the first two bytes.
 I2CBytes(S1, cmdbuf, num_of_bytes, outbuf);

 // now, outbuf contains any response from the slave, so we
 // can react to it here.
 Wait(10000);
}
```

On the Arduino side, we keep it simple for this demo as well. We set up the Arduino as a slave with the Wire library, with an address that matches what we use on the NXC side. We set up serial communications so we can observe what's happening as it happens. We set up two *handlers*, or functions that get called automatically when something happens. One is called when we receive data over I2C, and the other is called when we are requested over I2C to send data.

Here's the Arduino code:

```
#include <Wire.h>

#define SLAVE_ADDRESS    0x43
```

```
void setup()
{
 Serial.begin(115200);
 Wire.begin(SLAVE_ADDRESS);
 Wire.onRequest(requestEvent);
 Wire.onReceive(receiveEvent);
 Serial.println("Hello world.");
}

void loop(){
}

void requestEvent(){
 Serial.println("Data requested");
 Wire.write(0x01);
}

void receiveEvent(int bytesReceived){
 Serial.println("Received data:");
 for (int i = 0; i < bytesReceived; i++)
 {
 Serial.println(i, DEC);
 }
}
```

To run this demo, upload the Arduino code to your Arduino, and start the Serial Console at 115200. Upload the NXC code to the NXT, and navigate to the program using the NXT buttons. Run the program. When you do, you should see information about what's happening over I2C on the Serial Console. While this example isn't the most useful by itself, it shows that with just two wires you can connect the NXT to the Arduino, and use the vast ecosystem of Arduino examples and add-ons with your NXT, with your NXT being "in charge."

XBee Wireless Modules

XBee is the name for a series of radio modules (Figure 10-18) from Digi International. There are now a large number of modules sold under the XBee line, as well as other radio modules sold by other companies with the same form factor. A small disclaimer: because there are so many XBee options and only so much space we can devote to this topic, this summary only scratches the surface. If you're interested in this topic, Digi has been doing a great job providing hobbyist resources (*http://examples.digi.com* is a great place to start). There is also an excellent book, "Building Wireless Sensor Networks" by Rob Faludi that goes in depth with the Series 2 XBee modules.

XBees are used often in Arduino projects to wirelessly connect Arduinos to each other, or to a computer. While many computers have Bluetooth built in, it is not a standard option to have XBee support built in, so to connect to a computer using an XBee, you'll need a piece of hardware to do it. There are small standalone bridges that can connect an XBee to your computer over USB, and you can also use the official Arduino XBee Shield to connect an XBee directly to your computer.

Figure 10-18. *An XBee Wireless Card*

Connecting an XBee to an Arduino has two parts to it that may trip you up. First, the XBee pins are spaced at 2 mm between pins, instead of the bread-board-standard 2.54 mm. There are a variety of inexpensive breakout boards that have a 2 mm spaced socket for the XBee on top, and 2.54 mm pins out the bottom. Second, XBees run at 3.3 V, while most Arduinos run at 5 V. A bit of interface circuitry is required in order to communicate between the Arduino and the XBee. This can be done easily on a breadboard, or you can purchase an XBee adapter that provides a socket as well as allows 5 V interfacing. There are a variety of these available on the market, but throughout this book we've used the Arduino Wireless Shield, as it provides a socket, interface circuitry, and a switch that enables you to connect the XBee directly to your computer, all in a form-factor that directly plugs into the top of your Arduino. There is also an Arduino Wireless SD shield that provides all the same features as the Wireless Shield, but also includes a microSD slot. Much of the hardware to connect XBees and SD cards to an Arduino is the same, so this is an inexpensive upgrade to the shield that provides more functionality.

As mentioned previously, there are a vast number of XBee options. One of the main options is Series 1 or Series 2. Series 1 modules aren't actually labeled Series 1, but Series 2 modules do say "Series 2." Series 1 modules can talk with other Series 1 modules, and Series 2 modules can talk with other Series 2 modules. They cannot talk to each other. In each series, there are Pro and non-Pro options. The non-Pro options go around 100 feet indoors, and around 300 feet outdoors with line-of-sight. Pro options take more power, and have more range.

Each of these types has multiple antenna options. There are two options with built-in antennas: chip and whip. The chip antenna is lower profile, but doesn't have as much range as the whip antenna. The whip antenna modules have a small wire, about 1 inch long, which serves as the antenna. There are also options with antenna connectors to connect your own antenna. You can have a mixture of Pro and non-Pro on a network, as well as a mixture of different antenna types.

Series 1 XBees are configured at the factory to talk to any other Series 1 XBee that hasn't been reconfigured. The shipped configuration requires a serial connection at 9600 baud. In the simplest case, if you get two sets of an Arduino, a Wireless Shield, and a Series 1 XBee, and write two simple programs, one where it establishes a serial connection at 9600 baud, and then does a `Serial.print('a')` every second, and another program that toggles the LED on pin 13 whenever it receives a byte on the serial port, you can move them around and watch the LED on the second Arduino blink. If you turn the power off to the first Arduino, the LED will stop blinking, because it isn't receiving a transmission.

In the simplest mode, Series 1 XBees broadcast serial traffic to every other Series 1 XBee within range. This is how we've used them throughout this book. When we used multiple XBees, we distinguished them by prefixing the data with a special character. This was for simplicity. XBees are extremely configurable and are capable of many things! For instance, you can connect buttons and sensors directly to XBee pins, bypassing the Arduino completely. If you craft the packets yourself, or use a library that does so, instead of sending the XBee serial text directly, you can control the destination instead of broadcasting to everyone. You also can encrypt the network. If you make a network with Series 2 XBees, they can make a mesh network, which means that packets sent from one XBee to another will try to find their own way through intermediate XBees to reach each other. As you can tell, it can get very complicated very fast!

Epilogue

That's it, that's all we have. Working on this book has been an incredible experience that has tested us repeatedly over the past year. We learned more than we ever thought we would, and in turn have passed those discoveries on to you.

So, what's next? That's sort of up to you. We'd love to hear about your experiments with combining Lego and Arduino technologies. Send along your projects and we'll post them on Wayne & Layne!

Pleasant making!

John, Matthew, and Adam
Minneapolis & Pittsburgh
October 2012

Index

Symbols

K

Karvinen, Kimmo, 65
Karvinen, Tero, 65
Keytar, 203–236
  assembly instructions, 208–230
  build Lego model, 208–229
  electronics, 204–205
  install electronics, 229–230
  Lego elements, 206–207
  parts list, 204–207
  programming, 230–234
  tools, 204–205
Kinect, 66
KK series, 284, 291
knobs, 18–20
Krumpus, Michael, 55

L

L293D chip, 71
L293D half-bridges, 60
L293D H-bridge chip, 283, 284, 293
L293D motor control, 158
L298 full-bridge motor driver, 60
Ladyada, 66
Ladyada's Multimeter Tutorial, 139
lamp, 238–282
  assembly instructions, 243–275
  build Lego model, 244–270
  download and install app, 273–281
  electronics, 238–241
  install electronics, 271–273
  Lego elements, 241–243
  parts list, 238–243
  program, 275–281
  tools, 238–241
LCD screens, 27
LEDs, 34, 54, 142–143
  power, 54
left_speed command, 196
Lego
  modding, 70
  sourcing, 73
  System, 43
Lego Digital Designer, 190, 243
Lego Education, 40

Lego Group, 31, 32, 37
Lego Mindstorms, 3, 31, 32–39, 158, 241
  add-on electronics, 41–42
  Android app for, 238
  buying more, 40–41
  connectors, 37–40
  expanding, 39–40
  mechanics, 37–40
  motors, 3, 35, 35–37
  non-Mindstorms lego bricks, 43
  non-standard parts, 5
  NXT 2.0 set, 5, 31, 34
    Technic system, 37
  NXT brick, 33–34
  NXT set, 32
  NXT system, 145
  Power Functions, 44–45
  sensors, 3
  set, 73
  Technic beams, 37–40
  third-party bricks, 42–43
  wires, 37
Lego motor, 35, 39
Lego parlance, 35
Lego pneumatics set, 45
Lego robotics
  anatomy, 31–48
Lego's Pneumatics Add-On Set, 41
leJOS, 46
Leonardo, 26
libraries, 27–29
  Bricktronics, 27
  motor control, 27
  PS2 keyboard, 27
liftarm, 13–20
LiIon batteries, 297
Lilypad, 58
Linear Actuator, 45
LiPoly batteries, 297
lithium batteries, 297
logic HIGH (digital sensors), 146
logic LOW (digital sensors), 146
LoL Shield, 61
loop() function, 29, 90, 131, 193, 195, 200, 279

M

main loop, 27
MAKE, 67
MAKE: Arduino Bots and Gadgets, 65
MAKE ecosystem, 67
MAKE: Electronics, 152
MAKE Magazine, 64
MAKE: Projects, 67
Maker Shed, 57, 58, 59, 67, 152
  motor controler shields at, 293
  plate patterns available at, 288
  power supplies at, 295
Making Things Move, 66
Making Things See, 66
Making Things Talk, 66
Margolis, Michael, 62
MAX_BACKWARD command, 195
Maxbotix LV-EZ1, 150
MAX_FORWARD command, 195
MAX_PITCH command, 195
MCP23017 I/O port expander, 283, 284
mechanics, 37–40
microcontroller chip, 145
Microsoft, 66
microUSB cable, 26
Mindsensors Motor Multiplexer, 41
Mindstorms motors, 69, 71, 294
Mindstorms, sensors, 34–35
Mindstorms wires, 4, 20, 290
MIN_FREQUENCY command, 233
MIN_NOTE_DURATION command, 234
MIN_PITCH command, 195
minute_position command, 90
MIT Media Lab, 32
modding Lego, 70
Molex connectors, 191
  adding to LEGO wires, 290–291
Molex pins, 284
Molex P/N 09-50-8063, 284
Molex P/N 22-01-3047, 284
Motor Controller, 191
motors, 35–37, 291–294
  3D printing, 35
  DC, 35
  DC brushed, 291, 292–293
  driving DC brushed, 292–293

Colophon

The heading and cover font are BentonSans, the text font is Myriad Pro, and the code font is TheSansMonoCondensed.

About the Authors

John Baichtal is a contributor to *MAKE* magazine and *Wired*'s GeekDad blog. He is the co-author of *The Cult of Lego* (No Starch Press) and author of *Hack This: 24 Incredible Hackerspace Projects from the DIY Movement* (Que Publishing).

Matthew Beckler is a graduate student in electronic engineering at Carnegie Mellon University, and is a co-founder of Wayne and Layne, LLC, where he makes open source hardware.

Adam Wolf is a firmware engineer at an electronic design services company, and is also a co-founder of Wayne and Layne, LLC, maker of the Bricktronics Shield used in this book's projects.

Have it your way.

O'Reilly eBooks

- Lifetime access to the book when you buy through oreilly.com
- Provided in up to four, DRM-free file formats, for use on the devices of your choice: PDF, .epub, Kindle-compatible .mobi, and Android .apk
- Fully searchable, with copy-and-paste, and print functionality
- We also alert you when we've updated the files with corrections and additions.

oreilly.com/ebooks/

Safari Books Online

- Access the contents and quickly search over 7000 books on technology, business, and certification guides
- Learn from expert video tutorials, and explore thousands of hours of video on technology and design topics
- Download whole books or chapters in PDF format, at no extra cost, to print or read on the go
- Early access to books as they're being written
- Interact directly with authors of upcoming books
- Save up to 35% on O'Reilly print books

See the complete Safari Library at safari.oreilly.com

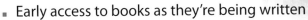

Get even more for your money.

Join the O'Reilly Community, and register the O'Reilly books you own. It's free, and you'll get:

- $4.99 ebook upgrade offer
- 40% upgrade offer on O'Reilly print books
- Membership discounts on books and events
- Free lifetime updates to ebooks and videos
- Multiple ebook formats, DRM FREE
- Participation in the O'Reilly community
- Newsletters
- Account management
- 100% Satisfaction Guarantee

Signing up is easy:

1. Go to: oreilly.com/go/register
2. Create an O'Reilly login.
3. Provide your address.
4. Register your books.

Note: English-language books only

To order books online:

oreilly.com/store

For questions about products or an order:

orders@oreilly.com

To sign up to get topic-specific email announcements and/or news about upcoming books, conferences, special offers, and new technologies:

elists@oreilly.com

For technical questions about book content:

booktech@oreilly.com

To submit new book proposals to our editors:

proposals@oreilly.com

O'Reilly books are available in multiple DRM-free ebook formats. For more information:

oreilly.com/ebooks

O'REILLY®

Spreading the knowledge of innovators

oreilly.com